ADVENTURES ADRIFT

By
Don Brown

Adventures Adrift

by Don Brown

Published by Pacific-Noir Press
1278 Glenneyre St.
Laguna Beach, CA 92651

Copyright © 2023 Pacific-Noir Press/Wendy Brown

All rights reserved. No portion of this book may be reproduced in any form without permission from the publisher, except as permitted by U.S. copyright law.

For permissions email: Pacific-Noir Press bylinecl@mac.com.

Design and Layout: Mike McCullen
With original art by Kevin Short and Martin Sugarman

ISBN: 978-0-9849504-7-8

First Edition

ABOUT ADVENTURES ADRIFT

The book you are now holding is a literary artifact, essentially faithful to the original text of Don Browns' hand-written manuscript.

Keep in mind that Don's manuscript was composed "then." Don's China, during the 1930s/'1940s, had been wracked by decades-long warfare. Subsequent revolution brought sweeping social and jolting political alignments that demanded "restructuring" to reflect their new "socialist cultural reality."

Such was the case in post-bellum China under the newly ascendent Communist government. Control of language became an important feature. In 1979, the People's Republic of China replaced the long-used and familiar 19th century Wade-Giles system of romanizing——i.e. transcribing, transliterating and pronouncing standard Mandarin Chinese into English using the familiar Roman alphabet.

Proper nouns denoting once-familiar cities, regions and geographic locations were arbitrarily reassigned the new Pin Yin notation; as a result they no longer correspond with today's maps and charts.

"Peking" became "Bejing." "Tinsing" became "Quing Dao."

We've included the original Wade-Giles spellings followed by the contemporary Pin Yin pronunciations.

Additionally, contemporary graphic and digital media communication styles differ. Once widely applied forms of diacritical punctuation——used as pronunciation keys in the transliteration of foreign languages in popular media are rarely used, if at all.

In an effort to facilitate following Don's wide-ranging sea-going career, additional page-below footnotes have been used to provide context items and events specific to Don's narrative.

Follow Don's wartime geographic locations along China's contemporary coastlines depicted by today's 21st Century cartography on page 122.

Craig Lockwood
Editor/Publisher

*I dedicate my father Don Brown's memoir, **Adventures Adrift**, to the enduring memories of two special people who meant so much to him, Li Teh Chuan, and Dr. James Robert Graham.*

These wonderful and beloved individuals shone as heroic bright stars in the universe of that tragic time and place. Our friend Craig Lockwood encouraged him to tell it.

— Wendy Brown-Barry

ACKNOWLEDGEMENTS

I would like to thank the following people for making this book a reality.

Lew Parlette for that first call and inquiry about my Dad's book.

My husband Kevin Barry, for his ongoing support, and for his understanding of how important it is to me to publish my Dad's story.

Kristi Brown, for her interest in her father-in-law's story. She stepped up with a second set of eyes to help in the editing process, and she gave us a beautiful place to stay overlooking my home town during our visits to Laguna Beach.

Dale Ghere, a retired Laguna Beach Lifeguard, teacher, and long time local and mentor to many, who referred me to my editor Craig Lockwood.

Craig Lockwood, also a Laguna Beach Lifeguard—knew and worked with my Dad during the time he was writing this book. Because of their close association, Craig adds a very personal touch to this book. I thank him also for his extensive research and pictures that depict the history of China during the time that my Dad was there serving in the U.S. Navy.

Michael McCullen; Pacific-Noir graphic designer. We gave Mike a very difficult job. He intuitively seemed to know what we wanted. He took a personal interest in my Dad's story, and genuinely respected our wishes. Thank you Mike, for your professionalism and perseverance.

CONTENTS

Preface 9
Introduction 17

Part 1

USS Pope DD225 Tsingtao
19 September, 1937 27

East China Sea Along China's Coast
October, 1937 31

Tsingdao, China
January, 1938 37

Tsingdao: Mission Compound
1200 Hrs. 41
1300 hrs. 43
1800 Hrs. 44
1900 hrs. 45

Mission: Hospital Ward
2200 hrs. 51

U.S.N. Submarine Base Supply Depot
0610 hrs. 53

Base Supply Depot
0700 hrs. 55

Number Three Beach
0930 Hrs. 57
1745 hrs. 61
1800 hrs. 63

Mission: Hospital Compound
1900 hrs. 65
2340 hrs. 66

USS Pope DD225, underway
0750 hrs. 69

Tientsin
1500 hrs. 71

Dockside: Tsingtao
1350 hrs. 73
0800 hrs. 75

Aboard ship, portside
0845 hrs. 79

Mission Hospital
1545 hrs. 81
0330 hrs. 87
05 November, 1938 88
0800 hrs. 89
2000 hrs. 90

Part II

"Congrats, Brown." 95

Aboard Charles F. Hughes
0840 Hrs. 97

Portfolio 39 99

300 miles south of Iceland 101
07 December, 1940 101

Bamboo Telegraph 103

Tug-of-War 107
1942 107
August 08-09, 1942 109

"Proceed independently to Samoa" 111
August, 1945 113
1730 hrs. 113
October—November 1945 117

Afterword 123

PREFACE

▍THREE A.M.

I'm putting pencil in hand to write about a near-legendary man, my father.

I grew up in the best of times on the corner of Avalon and Thalia Streets in the small art colony of Laguna Beach, California. It was the 1950s. The wars were over, and daddy was coming home. I was so excited because I scarcely knew my dad. He was a Navy man, often away for long periods of time. Being so young, I wasn't able to understand the calamitous events that took him away from us. So, my mom and I waited for his return in our little house at 449 Thalia.

And every night when mom came to tuck me in, I always ended my prayer with, "...And I hope my daddy comes home soon."

Our house was once a two-car garage my mom's father, Grandpa Hatch, bought for $300, during the 1920s. Grandpa turned it into a livable beach house. Mom and dad bought it from him for $5,000.00. That was my first and only Laguna Beach home just a few blocks from Coast Highway, and the beach at the end of Thalia Street was accessed by a dirt path.

Mom and I spent a lot of time there playing in the water and sand. And, above us was what locals called "Top-of-the-World," Laguna's steep ocean-facing hillsides covered in fragrant eucalyptus groves where my father's mother, sister, and brother lived.

I waited with anxious anticipation at the end of the Navy pier in San Diego Harbor looking out at the huge gray ships, waiting for my dad to appear. He had been on duty with the U.S. Pacific Fleet, and we hadn't seen him for almost two years!

As I watched, a handsome story-book prince stepped out of a cluster of Navy officers in his impeccable white service-dress uniform, and came striding toward us. When he saw me, he scooped me up and held me high over his head laughing and smiling.

Don Brown had returned to his family. And that's when our life together really began. He had changed. His hair had turned gray, and he had some lines on his face. War had taken its toll. But I thought he was the most handsome man I had ever seen, and he was, MY Dad!

In time I discovered that my father was a true hero, and had earned the Congressional Lifesaving Medal! I heard stories of how he had moved up through the ranks to a Naval officer, and became the Captain of an oceangoing tug, the *Squanto*.

I was one of the lucky ones whose father returned from the war, and I was so proud to be his daughter.

Dad continued to improve on our house. Mom had a green thumb, and grew lots of flowers. I was so happy, surrounded by their love in this secure, magical place.

My brother Bill arrived 1952, and dad taught us to swim, surf, and fish. He was a master boatbuilder, and took us on trips to Catalina Island to free dive and fish. The *AHSO* was a 24-ft. cabin cruiser with a flying bridge, a galley, sleeping quarters and a deck for fishing.

Dad taught me how to read the ocean and the waves by having me ride on his back holding onto his shoulders, as we slid off the reef at Thalia Street. When I was in high school he persuaded me to enter the Brooks Street bodysurfing contest, which I won two years in a row. I was shy as a young girl, and my dad gave me the confidence to try new things.

With his encouragement I started taking ballet lessons at Barbara Payton's Dance Studio in North Laguna, and learned to board surf. I remember his inspirational teachings as he held one hand in the air with his thumb and first two fingers touching while his wrist moved them back and forth as though he was about to taste a fine delicacy. Then he would say, "Wendy, the world is your oyster."

Dad also found the time to take me hiking. We'd drop down off the Top-of-the-World into Wood's Canyon and the Laguna Beach of his youth, where he showed me caves, Indian campgrounds, and large slabs of rock with sea fossils imbedded in them.

On one occasion we packed a lunch and had Mom drop us off out on El Toro Road, and we hiked from there all the way back to the Top of the World. It was a beautiful spring day. The wildflowers were in bloom and the creeks were running.

As we hiked he would share stories of his childhood, and how he brought Todo, a ground squirrel home. Todo had been washed out of his nest in a bad rain storm. By the time Dad got home, Todo had revived in the warmth of his pocket, and turned out to be a good pet.

Another special treat was "Father and Daughter Night Out." Dressed in our finest clothes, he would take me to movies, usually musicals, like the "King and I" and "Oklahoma". We also went to the Melody Theater for live performances on a round stage. These were special times that few if any of my girlfriends got to have with their fathers. These wonderful hours that my father spent with me are treasured and remembered.

As I entered my teenage years, life began to change. I became disenchanted with some of my classmates as they drifted away to find new friends, and I became lonely and depressed. One day I arrived home in tears after I had been bullied at school.

Dad was there for me and wanted to know if there was anything he could do to help.

"If I just had a horse to ride and be my friend, I don't need those A-holes at school."

I was born with a fascination for horses that was nurtured by my grandmother. Dad's mother

was a horsewoman, and she took me on adventures to guest ranches where I had some of my first horseback rides.

So on my thirteenth birthday I was given Stormy, a red sorrel quarter horse with a diamond on his forehead. My grandma bought him for $250.00, PROBLEM SOLVED, and off I rode into the hills behind Laguna.

Over the next five years Stormy became my emotional equalizer. To pay for his feed and stable expenses while in college, I took a summer job waiting tables. After graduation dad helped me trailer him to Mariposa in the Sierra Nevada foothills where I started teaching elementary school. I was able to find a place to rent on a ranch out of town where I could keep Stormy and be with him every day.

While I missed the beach and ocean a lot, the bond it had created between my father and I was strong, and has endured. Eventually I discovered other bodies of moving water, like the Merced River.

While my father and I shared an adventuresome spirit, dad's love of the open sea was very difficult to embrace. For me, being out on the ocean with no land in sight was very unsettling, especially when he expressed his desire to sail alone to Hawaii in a dory that he built.

In contrast, my seeking adventure centered on the High Sierra. So I began exploring the mountains on horseback, while my brother remained in

Southern California following in dad's iron man footsteps, becoming a Laguna Beach lifeguard as dad had been, winning the Brooks Street surfing contest, and paddleboard races in both Regional and National United States Lifesaving Association lifeguard competitions.

While dad was a tough act to follow, he always allowed me to follow my own path and find my own pearls. What we discovered over the years was the vastness that drew us to want to explore these different places became a bond that kept us very close.

When my adventures inspired me to explore the growing literary genre of cowboy poetry, mom and dad were there to both support and cheer me on. When I coordinated our local cowboy poetry gathering, my parents were always there in the front row, for seven years!

As I followed my bliss in Mariposa, dad was following his in Laguna. While working at the Newport Harbor Department and later in retirement, many young men were drawn to our house. They were interested in tapping into his knowledge of the ocean, whether it was navigation, boat building, lifeguarding, surfing, diving for abalone, or fishing.

A strong camaraderie was formed between these Laguna locals and my dad that still exists to this day, even though he is no longer with us. They have stayed in touch with me as well, and we have all kept the memory of my father alive over the years.

To add to his long list of accomplishments, dad was also a creative writer, penning and publishing stories of his adventures tuna fishing in Mexico. He also started writing a book about his war-time experiences in the Navy, and specifically his time in China. As with most military men who returned from the war, he didn't share much of this with me. But he did on occasion talk about it with his protégés.

One day I received a phone call from Lew "Punky" Parlette. He asked me if I had the manuscript about China that my dad wrote. Lew had become interested in that period of history and wanted to read Dad's story. I said I did and would look for it. Climbing up to our loft where the family archives are stored, I searched but couldn't find it. I told him I knew I had it, and would keep digging.

Several months passed, and then one day there it was; in the bottom drawer of my night stand. The yellowed pages in the old three ring binder were fragile and beginning to tear. There is a "right time" and place for everything, and I figured it was high time to read dad's manuscript. What an experience it was to hold that manuscript in my hands, and begin to read about a Don Brown that I had never known.

As I carefully turned the brittle pages I was filled with emotion and anticipation over what I was about to read—this deeply personal account of a young sailor before World War II and after the Great Depression, I became even closer to this man who was to become my father. His writing opened a window to the past that allowed me to better understand what it was like to be a Navy man.

His descriptions were rich and colorful. I was especially moved by his account of the end of the war in the Marshall Islands, after spending many years on the open seas fighting for our country.

As I read, gratitude welled up inside of me toward my father for his bravery, compassion, and love for the people of this world. I invite you to open the window and experience with him, his Adventures Adrift.

Wendy Brown
Mariposa California,
2021

DR GRAHAM

I first met Dr Graham in Tsingtao, back in 1937 and even though he was quite unassuming he made an impression on me. It wasn't his physical size because he was thin, light complectioned, about forty and wore glasses. A bald spot was emerging through his light brown hair. I guess this feeling I had about him was the thing he was about to do.

I was a ~~third class~~ quartermaster on an old four stack destroyer in the Asiatic fleet and the Japanese had just invaded China. As the captain had announced at quarters, the ships orders were to evacuate all American citizens who wished to leave Chinese soil. We would take them to the S.S. President Coolidge anchored off the bund in Shanghai. It was reasonably safe and Japan had declared her neutrality with the foreign governments having interests in China. For openers, all she wanted was China.

We went up the Hei Ho river to Tientsin and commenced taking aboard families from the American Legation in Peking and it was here we became aware of the determination of the Japanese government. ~~Around the docks~~ Their troops outnumbered us ten to one, while across the river a Japanese battalion pursued the remnants of a Chinese army. It was wide open country, freshly plowed, with ~~and~~ what appeared to be a patch of millet or barley here and

INTRODUCTION

Holding Don Brown's manuscript and reading his words—jotted well over half a century ago in pencil on ordinary three-ring, blue-ruled school notebook paper—feels like peering into a somehow familiar, but never-seen place—a "somewhere/sometime" about which you've heard but never visited.

"Somewhere" and "sometime" constitute the connective tissue of the "near-past"—a period of time during which people still living can recall and clearly recount events that occurred and which they may have personally witnessed or experienced during their lifetimes.

However, be advised: the "near-past" is an amorphous frontier, a shrinking fabric, a period of time whose length and duration is conferred on each generation and denoted by our human life's century-mark.

Few of us live much beyond one hundred years, and thus the shelf-life of personal, first-hand experiences—our actual living connection to prior events and individuals—is tenuous, especially without notation or recording.

Another compelling characteristic of this "near past" is the direct connection of person-to-event. You are "...shaking the hand that shook the hand."

Stories being recalled and recounted include personal emotional impact. With the story-teller's *proximity* and degree of *involvement* to the actual event—listeners are swept along, through history's numerous and complex currents.

And because these events were actually witnessed through the eyes of an individual personally carried along by those great historical rip-tides, their stories provide a strong, subjective level of authenticity.

I had the pleasure of hearing Don telling these stories first-hand during the mid-1960s, as he was writing them.

At that time, I was working part-time for the Times-Mirror subsidiary, the Daily Pilot, writing an occasional feature and a weekly surfing page. I'd also recently published short articles and fiction pieces in SURFER and SURFING magazines.

My full-time jobs, however, were as a Laguna Beach lifeguard, as well as a reserve Orange County deputy sheriff working *per diem* on the Orange County Harbor Patrol where Don Brown was my sergeant. And Don was himself, a former Laguna Beach lifeguard.

Those of us growing up in Laguna Beach during the first half of the 20th century, lived in a small, somewhat insular American community, one of a coastal string of California beach towns that defies easy characterization.

Yes, to the point of cliché, Laguna Beach has been described in the popular press and travel literature since the 1890s as an "Artist's colony"—a result of climate, light, and landscape's dramatic potential as well as the residence of a few, and frequent visitation of a number of recognized American Plein-Air painters, including William Wendt, Edgar Payne, Franz Bischoff, Anna Hills, Frank Cuprien, Granville Redmond, Guy Rose, and Marion and Elmer Wachtel.

This narrow creative strata was comfortably interwoven between Laguna's ordinary small-town citizens: art gallery and other business and shop owners, bakers, grocers, schoolteachers, librarians, nurses, doctors, lawyers, firefighters, lifeguards, and police officers, priests and pastors—regular Americans some of whom were also creatively involved while fulfilling all of the above job descriptions.

In addition, Laguna Beach starting the 1920s, with a population of less than three thousand was part of the hub around which two increasingly popular aquatic sports—spearfishing/free-diving and surfing—would both figure importantly in Don's life.

Born on 29 October, 1913, in Little Rock Arkansas, Don grew up from early childhood in that bygone Laguna, where he became familiar with neighbors, his friend's parents, and community members who spanned the town's creative and flexible artistic spectrum, including painters, sculptors, musicians, actors in film and stage, opera singers, dancers—all of whom were involved at some level in producing expressive works of art.

This included several nationally recognized best-selling authors such as John Weld, Richard Halliburton, and Dana Lamb, the adventure-writer, whom Don had known since his teen-age years, when Lamb served as Laguna Beach's first Chief Lifeguard.

Don Brown was a "local boy" growing up during the 1920s, when Laguna *was* credibly an "artist's colony." Young Don's earliest efforts at combing the beach led to a developing repertoire of ocean-skills that amplified as he grew older, starting first with learning to swim and handle surf, then mastering the preliminary skills required for abalone and lobster diving, spearfishing, hand-lining and pole fishing, and handling small surf-craft—light skiffs and heavier dories—and most importantly developing watermanship, the always critical range of skills and the reactive capacity to avoid sometimes disastrous situations and overturns.

Each youthful encounter with the Pacific Ocean brought either a hard lesson or the satisfaction of realizing an increasing level of comfort in mastering the numerous techniques needed to adeptly launch and row various craft back through Laguna's often unpredictable shore-break surf.

There was always bodysurfing, and by the 1920s, the new skill of surfboarding introduced by the Hawaiians, lifeguard George Freeth, and four-time Olympic medal–winning swimmer Duke Kahanamoku whose singlehanded surfboard rescue of eight drowning fishermen off a capsized boat at the Newport Jetty in June 14, 1925 when Don was 11 years old had captured national headlines and become the fabric of local legend.

Don Brown's first stint at public service came in 1934, when he served as a Laguna Beach lifeguard. He'd previously taken some preliminary requirements at Santa Ana High School, including a standard American Red Cross First Aid course, and the Red Cross's new basic Water Safety and lifesaving courses which had been recommended to him by Ed Hobert, another of Laguna Beach's early lifeguards.

By this point Don's levels of ocean-skills, his *watermanship*, were well-developed and well-recognized.

Characteristic of Laguna Beach's youth, then as now, kids who spent summer-after-summer on the beach continually noticed and casually evaluated their peers.

In this limited milieu there is a long-standing constant continuum of judgment, observed, but unstated. Despite no one consciously "keeping the score," every one of your peers "knows the score."

Watermanship is not a credential. There are no merit badges, diplomas, medals or patches. The lessons are hard-earned, never bestowed, acquired only by experience and otherwise unavailable.

Being a good waterman, at its deepest levels is much like being considered by others as a good swordsman—or being a good horseman or horsewoman. Exhibited is a form of accumulated knowledge and experiential wisdom. A skilled waterman knows and acts as much on "what and when" to *do* as "what and where" to "*avoid doing*."

As Don would recall hearing from a Hawaiian friend: "Bugga know already, he *neva'* assume."

And key to it all is the ability to do and act in the water with a high level of supple kinetic grace.

Fundamental to the concept of watermanship is an unconscious and organic respect for the inherent challenge and wide spectrum of risks the ocean provides.

This demands an intuitive understanding of the *limits* of personal physical ability, along with the capacity to accurately judge other individual's capacities.[1]

Don Brown displayed in youth a graceful mastery of ocean-skills, including the ability to read, know, and handle surf. This included a range of lifesaving

---

[1] See: *"A Brief History"* by Craig Lockwood Co-Chair Oral History Committee, Surfing Heritage Foundation: Published in *Santa Monica Pier Paddleboard Race & Ocean Festival Event Guide*, 2011; Santa Monica. *"The Whole Ocean Catalog, Sports Resource Guide for the Active Waterperson,"* Surfer Publications, 1986, p.16., Dana Point, Calif.

techniques acquired as a Laguna Beach lifeguard—training that would serve him well in saving others, throughout his life.

With the Great Depression in full dysfunction and America's economy in epic collapse, career opportunities for a recent high school graduate for whom four years of college was not an economic option, seemed narrowly limited.

Enlisting in the United States Navy in 1934, Don would rise to the Enlisted rank of CPO (Chief Petty Officer) and then under wartime conditions, with a commission, to Lieutenant. When World War II ended in 1945, Don returned to Orange County, eventually joining Orange County's Sheriff's Harbor Department, where he served as a Sergeant.

Don was a dedicated and voluminous reader and a born storyteller whose local nickname was "Windy." In his later years he honed those characteristics into becoming a writer, eventually publishing several very readable pieces of short fiction.

As Don's young deputy, with a monthly schedule requiring law-enforcement training, shooting range time and a mandatory patrol schedule and Don having known me since childhood, my assignment to his watch in Newport Harbor, or the then-under-construction harbor at Dana Point, eighteen miles south was always welcome.

Don preferred Newport's usually uneventful "cocktail hour" evenings "dog watch" or the late-night Graveyard watch.

Climbing aboard, Don would carefully arrange his gear, making his Stanley Thermos reachable. Piloting the Harbor Patrol's little grey fireboat/tug or one of the powerful Doris-hulled red launches we would patrol Newport Harbor's dark waters.

Cruising from the Harbor Department's base on Bayside Drive across from Balboa Island to the Turning Basin and the Commercial Docks with the cluster of Jimmy Mag's commercial abalone diver's skiffs, we'd thread through Lido and Linda Isles with an array of graceful yachts nestled between the now-silent homes of Hollywood's cinema royalty.

My contribution to the watch included a quick stop at the Balboa Pavilion's dock and a sprint for doughnuts at the shop opposite fun zone's the enormous Ferris wheel while Don stayed at the helm.

Back aboard Don would skillfully back our way out into the channel. Throttled down to the speed of a lap swimmer, we'd idle past John Wayne's famous yacht "Wild Goose," a converted 1942 Navy minesweeper. Clicking on

the cockpit's chart-lamp Don would produce a couple of paper cups, loosen the cap on his thermos and pour two steaming cups of strong coffee.

Reaching behind his seat for his briefcase, he'd pull out his worn three-ring binder.

"Made a few more notes."

In the dark and cramped wheelhouse, his pages illuminated by the dim glow of the panel's instruments, Don would slip on a pair of reading glasses, and turn over the tug's wheel to me

With the little tug's engine grinding a muted counterpoint to his voice, our horizon lifted, expanded beyond Newport Beach, California.

We were now a vessel-of-imagination, becoming the DD225 "Pope's" launch, evacuating as quietly and inconspicuously as possible an American missionary and his family, or an American diplomat or businessman, pushing against the tide and a huge river's flow, up-channel, sailing in recollection on a red-light returning course against the strong tidal flood of China's Yang-tze River, passing now under the dominating black silhouette of a quarter-mile long Japanese battleship, echoing across Time's un-passable barrier in the clear simple words of a man who lived and wrote the story you're now reading.

In Memorium,
Craig Lockwood
Laguna Beach, California

PART 1

U.S. ASIATIAC FLEET: NAVAL BASE TSINGTAO (TZING DAO), CHINA

USS POPE DD225 TSINGTAO (TZING DAO)

■ 19 SEPTEMBER, 1937

I was a twenty-one-year-old, two-stripe Quartermaster on the *USS Pope DD225*, an old Clemson-class, four-stack World War One vintage destroyer, deployed in the Asiatic Fleet.[2] When I first met Doctor Graham in the port city of Tzing Dao the Japanese had just invaded China.

Despite being quiet and unassuming, Dr. Graham left an impression. It wasn't his physical stature—he was thin, light—complexioned, balding, about forty, and wore glasses. Instead, my feelings immediately centered on what he was about to *do*.

Earlier, at quarters, Captain Couble, had announced our orders: we were to evacuate *all* American citizens wishing to leave China, and transporting them to the *S.S. President Coolidge* anchored off the Bund in Shanghai.[3]

---

[2] The Japanese invasion of Manchuria began on 18 September 1931, when the Kwantung Army of the Empire of Japan invaded Manchuria immediately following the Mukden Incident. By 1937, the *USS Pope DD225* was "showing the flag" off the East China coast during the summers, while spending winters in the Philippines engaging in division maneuvers. Increased tension on China's northern borders due to the Japanese invasion of Manchuria made it necessary for the *Pope* to evacuate Americans from northern Chinese ports such as Lao Yao and Quingdao to Shanghai beginning 19 September, 1937. From 15 July to 20 September, 1938, the *USS Pope*, cruised in Chinese waters off Quinhuangdao returned 05 June, 1939, with the South China Patrol force removing American consulates and nationals. The *Pope* was stationed off Shantou and Běidàihé Qū; during 14 June – 19 August, observing the Japanese Navy en-route to Swatow and the subsequent bombing and occupation of the city. The *Pope* remained in this area until her return to Manila on 12 October, for the Neutrality Patrol off the Philippines. *Pope* transferred to Division 59 of the Asiatic Fleet 06 May, 1940, and resumed patrolling off China, 11 May – 24 June. *Pope* returned to Manila in late June on neutrality duty and remained on station there until 11 December 1941, when she got underway for Balikpapan, Dutch East Indies.

[3] "Quarters" refers to the daily morning muster for each division, announced as "Quarters...Quarters... All hands to Quarters for muster, instruction, and inspection."

Our sea-detail was considered reasonably safe, as Japan had declared her neutrality with all foreign governments having interests in China. For openers, all Japan wanted was China.

We steamed up the Hei He River to Tienstsin (Tianzing) and commenced taking aboard families from the American Legation in Peking (Beijing). It was here we became aware of the Japanese government's determination. Their troops outnumbered us ten to one, while across the river a Japanese battalion pursued the remnants of a Chinese army. It was wide open country, freshly plowed, with what appeared to be patched of millet or barley[4] here and there.

---

[4] A Japanese Imperial Army Battalion at that time consisted of 1,100 men organized into 4 companies, each of 180 men, along with other units. Commanded by a Lieutenant-Colonel.

Through my long glass I could see the Chinese scurrying like little black dots, southward toward the sea. When a Japanese shell landed just right bodies flew through the air like chunks of mud—and lay still.

We evacuated our people to Shanghai and steamed north, returning to Tsingtao (Tzingdao) on the Shantung (Shandong) Peninsula, and always we were shadowed by an element of the Japanese fleet, their hulls low but their pagoda-like fighting tops cresting on the distant horizon. At night they closed in until their top hampers were silhouetted against the stars.

When we moored to the dock in Tsingtao (Tzingdao), the guns of the Japanese Army rumbled like approaching thunder behind the distant hills. Soft concussions rustled across the rice fields and gently rattled the shutters in the warehouse alongside of us.

And near the warehouse stood Doctor Graham. He was almost hidden by a bulk of missionaries, their wives and children and luggage. They were pleading with him and Graham listened attentively as their voices competed with the bulbous rubber horns of the rickshaw coolies, and the sing-song exclamations of the Chinese as they frantically loaded the costal sailing junks surrounding our ship.

"Christianity will be snuffed out like a candle," a churchman exclaimed, "Look what the Japanese did in Manchukuo! Not one Christian church survived! I tell you God will turn his face on this land!"

The missionary extended his arm toward Graham, and by long habit it was a motion full of grace and benediction, a direct conduit between God and his little people. "It's not safe!" he said. "Come with us while there is still time."

"But how about your medical supplies, your drugs and chemicals? With things the way they are the church may shut down the hospital."

"I guess I'll have to use herbs," Graham smiled. "I've always wanted to make a deeper study of them."

The conduit like arm lightly touched Graham's shoulder.

"Doctor, you'll be the only white Christian left in all of Tsingtao (Tzingdao)."

"I'll still have wireless radio contact with our American legation in Peking and Shanghai. And don't forget, I have an excellent Chinese staff to see me through."

Doctor Graham was growing impatient, as were the rest of us waiting to herd the group aboard. He glanced up at the bridge of the destroyer.

"The Captain up there is looking restless. Be gone with you!" He spiritedly shook their hands, shushing them on their way up onto the Pope's foredeck.

"This is not the end of the world, and I have work to do."

As we cast off our lines and backed into the stream, the old four stacker belched a great ball of black smoke. Captain Couble took a quick look as it settled upon a cluster of sampans sculling across the channel.

"Quartermaster! Tell the fire room to knock it off!"

I still had time to catch a glimpse of Doctor Graham. He was just a little guy who gave a quick wave and then disappeared into the growing crowd of Chinese milling around the dock.

EAST CHINA SEA ALONG CHINA'S COAST

OCTOBER, 1937

The following weeks were busy ones. As the Japanese leap frogged southward down the China Coast, we stayed one jump ahead of them, removing American personnel, and Union Oil Company officials who provided oil for the "lamps of China," as Pearl Buck used to put it.[5]

Their white fuel tanks were conspicuous in every major port facility.[6]

When the Nippones (Japanese) Army reached the outskirts of Shanghai it became more interesting. We anchored in the Whangpo River abeam of the Union Oil tanks to show the flag and give them protection.

Across the river lay Shanghai proper, with its hotels, banking establishments, dog racing park, legations, international settlements and Blood Alley with its "fun" houses.

This was a seaport to satisfy any desire. Unfortunately, lying off the Bund, a thousand yards away, was the *Idzumo*, an ancient coal-burning Japanese

---

[5] While the author attributes this phrase to Pearl S. Buck, author of *The Good Earth*, 1932 Pulitzer Prize, and 1938 Nobel Prize winner, it was author Alice Tisdale Hobart who first used the term "*Oil for the Lamps of China*" as the title of her 1934 bestseller. Hobart's husband was a Standard Oil Company executive. Hobart's novel describes a young executive in an American oil company in China from the early 1900s through the Nationalist Revolution of the 1920s. In the 1890s Standard Oil marketed kerosene to China as lamp fuel, adopting and trademarking the name *Mei Foo*, the name of the tin lamp that Standard Oil produced and gave away or sold cheaply to Chinese farmers.

[6] During the 1890s, Standard Oil began marketing kerosene to China's large population of close to 400 million as lamp fuel. For its Chinese trademark and brand, Standard Oil adopted the name Mei Foo which also became the name of the tin lamp that Standard Oil produced and gave away or sold cheaply to Chinese farmers, encouraging them to switch from vegetable oil to kerosene. Prior to Pearl Harbor, Standard was the largest single U.S. investment in Southeast Asia.

battleship with a ram-like bow, looking as though she had sailed out of the pages of a history book.

Her main battery consisted of twelve-inch guns (against our four-inch) and we lay there and looked at each other.[7] When she manned her battle stations we went to general quarters[8] and sometimes visa versa. When she broke out the rice and chop sticks, we chowed down on sauer-kraut, boiled potatoes and weenies.

I guess she felt the odds were against her because eventually a sleek Japanese heavy cruiser with two destroyer escorts came up the Whangpou and anchored even closer.

We were staring right down their throats into the bores of those enormous twelve- and eight-inch rifles and it was impressive.

We kept thinking about the *U.S.S. Panay*, our gunboat that had been sunk on the Yangtze a short time before. It was a case of "mistaken identity" of course. The Nipponese (Japanese) government had made their apologies. If anything happened here it would be over in seconds with our best Sunday colors flying.

As Japanese troops entered the sprawling native district around Soochow Creek, the ships opened up with a walking barrage.[9] Buildings exploded, pieces of sampans flew through the air and soon all of the wooden structures were on fire. It was unopposed mayhem.

For the Chinese it was business as usual. Since the opium wars of 1840, these sociable, accommodating people had become inured to the sound of foreign guns and troops tramping across their lands, giving up a little piece here, a city there, as though it were a finger or a toe and surely no worse than famine or plague.

After the Northern sector was softened with heavy artillery shelling, the *Idzumo* and the cruiser swung their guns across the central part of the city, past the international settlements of the British, Russians, French,

---

[7] Brown's ship, *DD225 "Pope"* carried four 4-inch (100 mm) guns; One 3-inch (76 mm) gun; Two M-1919 .30 cal. 7.62 mm. belt-fed machine guns, and 12 x 21-inch (533 mm) torpedo tubes.

[8] General Quarters, "GQ" is a U.S. Navy term announcing to a warship's crew to immediately assume assigned battle stations and positions becoming combat-ready.

[9] A deadly form of extended creeping artillery fire along a pre-determined advancing line.

and Americans until they were bearing on the south side of the Whangpoo (Huangpu) in the general direction of the white Standard oil tanks.

And of course, we were blocking their view.

At such short range their guns were depressed until we could almost look down their muzzles. Our four-inch guns looked like pea shooters. We felt like David looking at Goliath. Like threatening an elephant with an air rifle, and so Captain Couble unlimbered the torpedo tubes. While Diz, the chief torpedo man cranked them around, the Idzumo let a salvo go. As she disappeared behind a sheet of smoke and flame, the concussion from the guns hit us. The blast of hot air smelled of cordite and gun cotton while the twelve inch shells sailed down our side like pickle barrels at mast height level. For a moment no one was sure, but her true point of aim soon became obvious. Chinese troops had decided the safest place to be was the fuel tanks, with us out there protecting them, and they were making a run for them.

We plugged our ears and ate gun cotton all afternoon and then near sunset a Union oil tanker came up the channel.[10]

That Merchant Marine skipper displayed a lot of cool. Lying off for twenty minutes studying the empty dock waiting for him at the Union Oil tank farm, he judged the current, and waited out the target practice.

And then, after we trained our torpedo tubes directly at the *Idzumo* and the cruiser, he moved slowly into the dock. He hoisted "Baker"—the international red flag signifying, **"I am engaged in a dangerous operation,"** to the tanker's yard arm.

The tank farm was soon hidden behind the empty tanker and the Japanese ships ceased fire. They put a landing party ashore and surrounded the tanks while the merchant ship connected her hoses and sucked the juice out of them.

10 Guncotton is a highly flammable high-explosive compound developed in the late 1890s, as a gunpowder replacement in artillery ammunition.

> ## *"Someday we'll have to kick the shit out of those guys."*

Why those oil-storage tanks didn't go up in flames remains a mystery. I guess it was the will of Allah, or Confucius, or God, or perhaps all three of them.

The following morning, however, the tanks were drained dry, as we got underway. Living became a little rougher for the folks remaining in the elite International settlements of Shanghai.

We provided the tanker an escort to the China Sea and she sailed with the largest Chinese crew ever to man a merchant ship.

As we understood it, her captain formally signed *all* of the Standard Oil Company's employees on as her "crew" in order to meet international requirements. About two hundred of the company's staff had made it to the tank farm.[11]

As we steamed by the *Idzumo* the usual courtesies extended to foreign naval vessels were over. No more whistle signals, no more righthand salutes, no more dipping of colors.

As we stood by our guns and our ancient torpedos, Captain Couble peered at the battlewagon with his binoculars and growled out of the corner of his mouth, "Someday we'll have to kick the shit out of those guys."

---

11 We were later informed that the tanker's skipper sold the oil in Manila while the "excess crew" made their way through the city, scattered out through the cane fields, mahogany forests, and plantations in the Philippine Archepelago.

TSINGDAO (QUINGDAO), CHINA

■ JANUARY, 1938

We didn't return to Tsingdao (Quingdao) for several months and by that time Dr. Graham had become a distant memory.

Shantung (Shandong) province had no international settlements of importance, no foreign concessions that concerned us except for the U.S. submarine/naval base at Tsingdao (Quingdao). To avoid any friction or surprises by the Japanese, the U.S. was preparing to pull back to Cavite, in Manilla Bay.

The base still had a skeleton force, and we had been assigned to a mail run. Besides the usual sacks of mail, we had a half-dozen boxes addressed to Dr. Graham. I found out later, that they contained medical supplies, such as iodine, swabs, soap, bandages, instruments, ether, morphine, and the commonly used topical painkiller of that period, cocaine.

Things had settled down a bit, and now that the Japs were in, the various foreign countries were getting used to them. We were looking forward to the Russian cabarets, the daughters of exiles that had escaped to China during the Russian Revolution. There were dozens of them lined up along the wall across the dance floor. Long-stemmed beauties that melted into your arms and we were heroes in tailor made blues with a pocket full of mex, Chinese dollars, that lasted forever at ten cents a dance, and if you scored you took them home. *Dear God! I wonder whatever happened to them.*

When we pulled into the dock at Tsingdao (Quingdao) I had managed a two-day pass, but first there were things to do, clean up the bridge, stow loose gear, and by the time I hit the dock Doctor Graham was there. He was silently struggling with his boxes, tilting them into a bullock cart. I guessed he couldn't

afford a coolie so I gave him a hand. He didn't know anything about the military at that time, and kept calling me "Sir!"

Me? A third class quartermaster?

"Thank you, Sir" he said. "Can I give you a lift anywhere?"

I looked at the flat board used for a seat and replied, "No thanks Doctor, I'll use a rickshaw."

Although my mind was on other things, there was something about Doctor Graham that hooked me. He was a positive sort of guy, an outgoing American who probably got things done, whatever those were, with good will and a smile.

"Well, OK," I said. "How far are you going?"

"Anywhere you want to go, Sir," he replied.

"O.K., *where* are you going?" I asked.

"To a compound about five kilometers out of town."

"O.K. Let's go," I heard myself saying. "I'll help you unload." After all, I could walk back to town in an hour.

Doctor Graham aimed the cart at the rim of the sun hanging in a smoky haze on the west side of town. We turned this way and that through the narrow cobblestone streets, and it was the usual scene.

Shops and stalls locked elbows on every block in one great continuous garage sale. Here all the needs of man could be found, except perhaps eternal youth. And even then there were, for the older men, elixirs and herbs and aphrodisiacs of ground reindeer horn from Mongolia and Siberia.

There were brass pots of all sizes, pewter ware, clay water jugs, and blown glass goblets, jars of pickled eggs, very ripe with a dark blue, purple color. Hanging, were smoked Peking ducks with a brown glaze. We called them shellacked ducks, and they were very edible. An infinite variety of everything—

and through it all were the sounds of the vendors enticing the buyers—the noodle man beating his gong and singing out *"Chieh Mien! Chieh Mien!"*

The vegetable vendor with his cabbages, bamboo shoots, and beans, shaking his rack of swinging gongs as though it were Chinese New Year, the sesame-cake man taping his drum and calling *"Chao Ping!, Chao Ping!"* And through it all were the earthly smells of a Chinese city, of burning charcoal, garlic, and urine.

Perched on the bullock cart we went through the streets, ducking our heads beneath the red and green paper dragons hanging over us and it seemed that half the people in town knew Doctor Graham.

"Hau-bu-hau! Doctor Graham" they would call and wave.

"Hau, Mister Lee," he would reply, and then they would ramble off into a dialogue that lost me.

We started pulling up the hill leading out of town, past the cabarets and assorted bars. It wasn't like it used to be. It was quiet now, with a few Japanese troops hanging around wondering where the action was. They were on liberty the same as I, and they starred unbelievingly as we ambled by. The white skinned big noses were still around.

"Have they been leaving you alone?" I asked.

"Not bad at all," Graham replied. "General Itagake and some of his staff have been around a couple of times. But so far they've left me alone." [12]

"A General? That's a lot of horsepower!"

"Yes, he's the Commanding General in Shantung (Shandong) Province, speaks pretty good English too." He turned and smiled. "I say Sir, I'm glad you came along."

Graham was fifteen years my Senior, so I replied, "I'm an enlisted man, Doctor. No need to call me 'Sir.' Just call me Don, Don Brown."

"Any young fellow out here representing his country deserves to be called Sir" he replied "and it's plain to see you're a petty officer[13] and a gentleman. However, if you wish, let's make it Don."

---

[12] General Itagake was considered primarily responsible for the 1931 Mukden Incident in Manchuria. He was captured by the Allies and in 1948, tried and executed for war crimes.

[13] "Petty Officer" is a corruption of the French "petite" or smaller officers—a military term for NCOs.

TSINGDAO (QUINGDAO): MISSION COMPOUND

▍1200 HRS.

Graham's compound was a substantial affair, set on the side of a hill where the earth consisted of hard shale and was unusable for farming. The main building was of brick with a tile roof—and fanning out from it like the spokes on the hub on a wheel—were the hospital wards, low barracks-like structures with whitewashed walls and thatched roofs.

The church, sitting off to one side, had a bell tower. As with all important buildings in China, it had curved-up eaves to keep evil spirits from sliding down the roofs into the building contaminating the whole cotton pickin' shebang. I pointed at the eaves and said, "Are you sure Christians built this church?"

Graham laughed, "The Chinese are a very practical people. They like everything going for them, Confucius, Buddha, Jesus, and no matter how you cut it, I'm for them."

"It's doubtful they would have gone in there for serious worship without those curved eaves. Why, if I had my way," he continued, "I'd put a prayer wheel outside if it would help them any. It's heart that counts and they have plenty of it."

A seven-foot wall of adobe and mud surrounded the buildings. Just a few years before it was a vital necessity for every small village, a defensive buttress against the warlords and brigands that raided the countryside. They were still at work farther inland, fighting the Japanese under the pay of Chiang Kai-Shek and someone higher up north in Hopei Province, someone by the name of Mao.

About twenty farming people in baggy pants and quilted cotton coats were waiting for Doctor Graham, seated on the steps and benches going into

It was a bad break with a bit of bone protruding through the skin.

the main building. As we entered, they stood up and quietly stared. Not the Chinese staff however. They immediately began ushering everyone into the main hallway.

"It looks like you have company, Doctor," I said. "I'll take care of the cargo. You go ahead."

I managed the whole operation without getting a spot on my tailor-made blues, and when I went in to say good-bye things were fairly under control. Half of the waiting Chinese were family. Kinfolk who had brought their casualties to Doctor Graham, and they were all serious. As the doctor told me later, they still preferred to treat their own with poultices, herbs, and acupuncture. When their traditional remedies failed, he got the desperate ones that could go terminal.

He was already working on a little girl with a broken leg. It was a bad break with a bit of bone protruding through the skin. Her brown little shoe-button eyes were bright, but she was looking into our foreign faces and she was frightened.

Doctor Graham and the nurse spoke soothingly to her in Chinese while I found myself stroking her hair saying, "You'll be O.K. Honey. Hau! Doctor Graham is number one."

And he was, quick, efficient, sing-songing away in Chinese dialect to his staff, while preparing the most serious cases. As he finished up with the plaster cast, he raised his eyes.

"Why don't you stick around? If you're any good at first aid—we can use you later on."

"I can manage," I said.

"Good" he replied. "Wash up and get a mask. We're going to do an appendectomy."

▌1300 HRS.

Doctor Graham was tireless, probably because he was dedicated to his work. There was no x-ray machine, and only a very small lab out of necessity. He was thoroughly acquainted with the human machine, with ruptured spleens, gallstones, and malfunctioning livers. The last patient had a bullet wound through his lower intestine. He was in his middle twenties, and his name was Yu Fong, a farmer from the next village.

"How did it happen?" Graham asked.

"I shot myself while cleaning my rifle."

After Yu Fong was under the ether and Graham had gone through the muscle into the abdominal cavity, I asked.

"Do you think that's true?"

"Probably not," Graham replied. "These people can't afford a gun. By the size of the hole it looks like a military round of some kind."

"Yeah, I replied. "A Jap rifle."

"Could be. This lad might be in a guerilla band—or the Chinese Army."

"It's probably best of we went along with the farmer routine," I said.

"Absolutely!" Graham replied. "But any young fellow who can walk in here with a hole through him like this, I'm not going to stand here and let him tip over."

"Do you think he has a chance?"

"Of course, If I can clean him up right. Peritonitis is the big thing here."

He turned to the Chinese nurse, and for the first time spoke to her in English. "We'll write this up as a gall bladder operation, Li Teh."

"Of course, Doctor," she replied in perfect English.

There was an attractive lilt to her voice as though she were still speaking Chinese. I stared at her. I'm sure my jaw dropped open.

For four hours we had been rubbing elbows while I played flunky, passing swabs, checking the sterilizer, and anything else I could see doing that would take the pressure off of them.

On occasion our hands almost touched while I passed her the surgical instruments, playing it sanitary and attempting to maintain a germ-free area. When she saw my clumsy efforts, her very dark brown eyes smiled at me above the gauze mask. They they seemed to have an honest, trusting look, as though we were man-to-man, people bound together here in a common cause.

1800 HRS.

By the time we were finished evening twilight had silently settled around us and while the Coleman lanterns and kerosene lamps were lit, I asked, "Did I hear right Doctor?"

"You sure did," Graham replied. "Li Teh, this is Don Brown. This is Li Teh-Chuan."

"Hi, Li Teh," I said.

She removed her mask and laughed. Her teeth were even and white.

"Hi?" she inquired. "That's Japanese for hello. You're not Japanese."

"That's American slang for hello," I said. "The Japs must have stolen it from us."

As we cleaned up, Graham continued, "Other than myself and my good brethren that have left, you are the first American Li Teh has met, and probably the first sailor ever."

"Well," I said hesitantly. "I'll try to set an example for them." I turned to Li Teh, "Where did you learn to speak English?"

Before she could reply, Graham cut in, "Let's check the ward and then have dinner. Will you stay with us?"

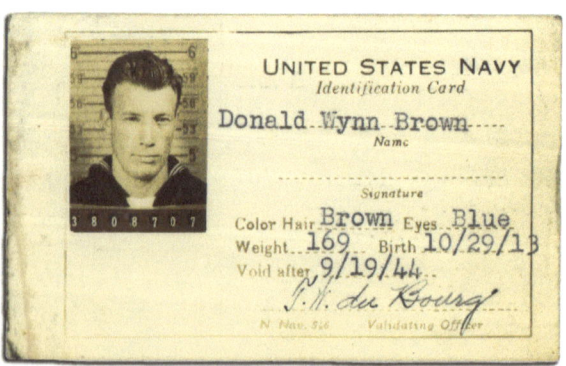

Small caution signs were flickering across my mind. Involvement can lead to commitments, to obligations, and a destroyer sailor in these times was dedicated to one thing, his vessel. Still, the cabaret had lost its appeal for me. As we walked down the corridor, Li Teh seemed to float along in soft soled slippers. She had an erect, fluid grace and in her English slacks and blouse, it was obvious she had a figure.

The staff ate together in a screened wing that was a part of one ward. Large Chinese woks were brought to a searing heat on charcoal braziers. A bit of peanut oil was added that popped around like raindrops on a red-hot griddle. Thin slices of lean meat were added, followed by chopped vegetables, bean sprouts, mustard greens, bamboo shoots, and anything else available that was in season. This was served over a bowl of rice with soy sauce and fish paste. It was a standardized nourishing dish, washed down with large pots of tea.

Now the room was transformed, taking on the character of a Chinese restaurant, noisy, bubbling with talk and laughter.

And it was here I became involved in the lives of Doctor Graham and Li Teh-Chuan.

1900 HRS.

Seated, with our meal Graham explained he had come to China in 1921, after two years of post graduate studies and doing emergency surgery in a general hospital in New York.

"It was a good place to learn," he said. "And even then, I knew I was going to come to China," he laughed. "I guess I can blame it on my grandfather. He was a sea captain, and when I was young he brought me a model of a Chinese sailing junk and told me sea stories."

Then he grew more serious.

"It was almost a compulsion that I come out here and I've never regretted it. These people are very basic, at least at least the ones I deal with. They look at life the way it *really* is with no hang ups. As a nation they are the most rational even tempered, happiest people in the world. I guess that's why countries with an axe to grind can come in here and take a slice of them.

"Li Teh-Chuan came to the mission in nineteen-thirty. She was a giveaway girl of fourteen and her parents felt it was time for her to be on her own. Still, she must have been something special to them because they gave her to the nearest Christian mission to work and to learn."

"It could have been worse, and it was here she was taught and learned English, Christianity, the history of her own country and foreign ways. And because she was intelligent, her mind expanded and blossomed along with her body, and she became a beautiful girl with the oval shaped face and slender figure of many northern Chinese."

"After three years with the mission when she was seventeen, fate once more stepped in and took a hand, and fate came in the name of General Chang Tsung-Chang, the warlord of Shandong Province. His troops, gorillas, some people called them, numbered in the thousands, and they had to control the peasantry in an area the size of Texas."

"General Chang collected the taxes from something like twenty million people! And he lived richly in a style equal to a king."

"As history records, Chang came up the hard way. Standing over six feet tall with his boots off, he started as a coolie loading ships from Hong Kong to Tianjin. Now, he was General Chang Tsung-Chang with over thirty concubines."

"The Monster," the local people called him and the Chinese who loved to gossip called him "Old Fifty Six," because he had devised that many methods of performing the sex act. But back in 1933 "Old Fifty Six" was having a little problem. Chiang Kai-Shek was reaching out from Nanking and putting on the heat."[14]

Chiang wanted to unify China and he marched into Shantung Province to argue the point. The War Lord's troops were getting shot up and he turned to Doctor Graham.

Riding in on horseback with a wagon load of injured, he said, "Fix them up." He congratulated Graham for his outstanding facilities, and when he saw Li Teh he went over her with a practiced eye. He opened negotiations to buy her. The bidding went up until Li Teh became the equivalent price of the entire compound. When Graham was forced to say she was not for sale, General Chang Tsung-Chiang flatly stated he would burn the establishment to the ground and he got her for nothing. Li Teh kowtowed before him in ancient

---

[14] Chang Kai-Shek was the protégé of a remarkable Western-educated physician/revolutionary, Dr. Sun Yat-sen, a political cult-like figure who wanted to unify China from the Quing/Manchu Dynasty. Chang Kai-Shek started as a revolutionary under Sun Yat-Sen, becoming a military and political leader of the then-newly formed Republic of China between 1928 and 1975, first on mainland China until 1949, when the Communists under Mao Zedong seized power, and afterwards, retreating to the offshore island of Taiwan.

Chinese fashion with her forehead and arms upon the floor, and he took her away that night.

"While "Old Fifty Six" was busy with Chiang, Li Teh was taken to his famous household.

"She was bathed and perfumed while the concubines advised her of the various sexual intimacies her master desired. She was given a palatial room, the likes of which she had never seen before. And she was cautioned that those girls who had refused to become a proper consort faced disfigurement, or having their throats cut."

> *He got her for nothing.*

She lived in Chiang's palace for eight moons, but as Doctor Graham added, "He wasn't around too much. If he wasn't busy defending his southern boundaries he was fighting Mao Tse Tung up North. It was a time of big decisions for the War Lords of China, "Which side? Which side?"

"But before Chang could make up his mind, he was assassinated by a hatchet-man in Tsinan."

"Li Teh-Chuan returned to the mission hospital and everything was the same, except for one thing. She had been the concubine of a great War Lord. For the Chinese with whom she worked, she had much face now, and almost unconsciously they set her up as though she were a minor bronze statue in a Josh House. They set her apart, and then they really began to notice she was just a little different."

"At times she wore foreign clothes, spoke in a foreign tongue, and carried herself as though she were a Manchu Princess. None of the men could possibly consider her in marriage. It was a creeping sort of thing that Li TEH didn't notice until a barrier was there.

She tried to break it down, but the feelings of warmth that once flowed to her from her own people was now no more than a trickle.

Doctor Graham sighed, "When I saw what was happening, I tried to fill the vacuum, and she became like a daughter to me." He smiled, "I dare say, she's the best RN in China."

I thought of my clumsy attempts to assist in surgery and began to apologize.

"I thought you did very well," he cut in. "You have a natural ability that will take you anywhere you choose."

I shifted the conversation to a more substantial subject, to logistics and supplies. "You have quite an establishment here. It's going to take a lot to keep it going."

"We're in good shape right now," he replied. "The shipment we received today should last six months. Of course I have another order in," and then his face lit up. "I believe the Chinese have really accepted me since the Japanese moved in. We've been living off the local people for over a month! They just keep bringing in whatever they can spare, and come to think of it, it's not so local either. We received six sacks of rice from Chiao Hsien, thirty kilometers away! But we may run out of warm clothes for the children you see running about. Our coldest month is still ahead and, all they get are made over hand me downs that are pretty threadbare."

And then a light bulb turned on in my head. A flash of genius, I felt at the time, and I said,

"The fishermen on the junks up in Chefoo made pants and jackets out of our Navy wool blankets. They were always trading us out of our old ones about this time of year, how about some of them?"

"Never look a gift horse in the mouth!" Graham laughed. "I'll take all I can get."

Other than the two blankets on my bunk back on the ship, the only other possibility was the submarine base which had almost been reduced to caretaker status. A small detachment of our Marines had been sent down from Tientsin to guard the place.

"Tomorrow," I said, "I'll see what I can do." And then I took the bull by the horns. As Graham had said before, it was almost a compulsion.

"Can Li Teh go with me? Perhaps we'll have time to go to the beach and catch the last of the summer sun."

Who was I kidding? It was November! But a warm October sun had still trailed on into the month and there was a chance the sand was still warm by the sea.

"If she wishes," Graham said.

Li Teh looked at me with those direct brown eyes, smiled and nodded. I felt like leaping into the air and kicking my heels.

And so it was arranged. Graham offered the mission rickshaw as though he was a future father-in-law offering the family Cadillac. I doubt if Graham ever used it.

MISSION: HOSPITAL WARD

2200 HRS.

That night I slept on a cot at the end of the clinic's ward listening to the groans of the patients as they turned.

Several times I walked down the row of beds barefoot and in my skivvies to assist in any way I could. And every time Yu Fong, the young farmer who had shot himself, followed me with his eyes.

He was out of the ether, and with no pain killers he was hurting bad. His forehead was moist and clammy, his lips parched and dry.

"Water," he whispered in Chinese.

I pumped a pan of water from the cistern, wiped his forehead, moistened his lips, and nodded encouragingly until he closed his eyes. With a bullet hole through the lower intestines, you just don't get water, at least not for awhile.

And before daylight I knew he wasn't a farmer.

Somewhere in the background, off in another room, somewhere I could hear the clacking of a treadle. Someone was weaving or making a carpet. It would start and stop, and then run along for perhaps a minute until I decided it was a foot-powered sewing machine, but at this time of night?

I listened to the groans, along with a young Chinese boy of about sixteen who could use a little more first aid, until finally daylight came.

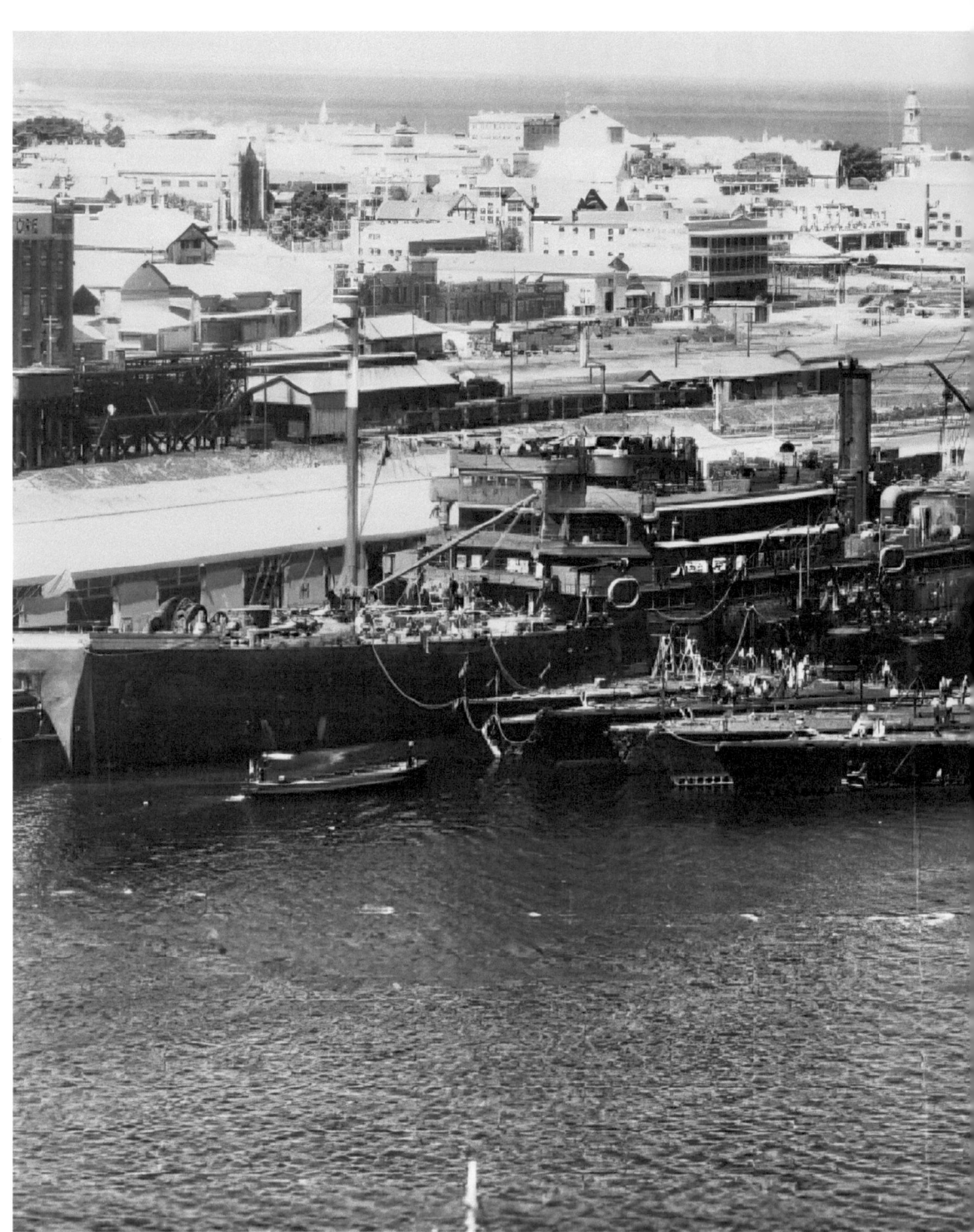

U.S.N. SUBMARINE BASE SUPPLY DEPOT

■ 0610 HRS.

For breakfast we had a warm thin soup, almost like bullion, topped off with a bowl of rice mixed with thin slivers of pork mixed with it, and by that time our rickshaw was ready.

It was a magnificent affair, an old fashioned model that had seen many years and it was in perfect repair. Comparing it with the modern city rickshaw was like comparing a restored Duesenberg with a new Chevrolet, the kind you walk around and say, "They just don't build them like that anymore."

For one thing it had larger, light-weight wire wheels with rubber tires. The seat inside the black lacquered cab was of a fine grain leather. It had a rubberized fabric top with lace curtains on the sides, and it sported a pair of polished brass carbide lamps, one on each black enameled fender.

The shafts for the coolie that pulled it were long whippy affairs of polished wood with a crossbar at the far end. This was lightly shaped and designed to resist the strength of a man's arms, rather than a yoke for an ox, or a team of horses. It had an elegant, ethereal quality about it, as though you could put it on a string and fly it like a kite and like every Asiatic sailor, I felt I was a connoisseur.

The coolie that pulled it was aging and matched the rickshaw. A leather cap with a small visor covered his graying hair. A pair of shorts and a jacket open to the waist covered his tall, sinewy body.

He reminded me of some old retainer kept on by a wealthy family in Newport, Rhode Island. But this old boy wasn't polishing a Cadillac. He was pulling a rickshaw in China. He was called upon to do his duty and he was proud, erect, and smiling. With his limited knowledge of the world, this was

[53]

going to be a big day for him. He was going to pull a foreigner, an American Warrior, a Big Shot, and in the seat beside him would be Li Teh-Chuan.

He took his position between the shafts while Li Teh stepped aboard. She wore a blue, full length Chinese dress, slit on the sides one inch above the knee. After I sat down beside her, he shifted along the shafts, adjusting his weight to ours. While the mission's staff waved, he honked the horn, rang a bell with his thumb and we were away.

Going into town Li Teh made it easy for me, answering my questions and in turn asking about America. I could feel her thigh pressing against mine, her knee against my leg. There was no room to get away. I let her study me and wondered how it would feel for a Chinese girl riding with an American sailor.

When I did turn and look into her eyes, they were untroubled and smiling.

BASE SUPPLY DEPOT

0700 HRS.

Li Teh was probably more relaxed than I because I had a little problem. How was I going to get those wool blankets?

Should I go through legitimate channels at the supply depot on the base? No way! Even if a supply officer wanted to help the mission, he would have to seek permission from higher authority, which could take weeks, and I wanted the blankets today.

Mulling it over, I watched the rickshaw coolie. He trotted at a steady even gait, pulling us effortlessly. Li Teh told me his name which I immediately forgot. Like all foreigners, I called the rickshaw boys "Chop Chop. Hey, Chop Chop! take me to Bubbling Well Road, Eh?"

Riding a rickshaw pulled by another human being was an accepted aspect of China's culture that I could never get used to. The Chop Chops of China considered it an honorable profession, and took much pride in it, especially the stronger ones. Li TEH explained it to me until finally I asked, "I want to try it. I want to pull you. Do you think he would care?"

"He would loose face!" she exclaimed. "He would be ashamed."

OK, then," I responded, "I want to try it. I want to pull you. Do you think he would care?"

"He would lose face!" she exclaimed. The indignity in her voice was forceful. "You must understand—he would be ashamed."

"Okay, we'll let it rest awhile," I said. "But start putting it to him easy, like I want to try pulling the rickshaw, just to see how it feels?"

We took a turn around the perimeter of the supply depot and the barracks. Except for a group of Chinese near the laundry and the Marine corporal standing duty at the main gate, it looked like a Nevada ghost town. He was standing at ease with a fixed bayonet on his Model 1903 Springfield rifle.

"Oh!" Li Teh said excitedly. "Is this where we get the blankets?"

I decided to use the positive approach, and as we pulled halfway through the gate the corporal came to attention with non-committal eyes on me.

I flashed my I.D. card as though it were direct orders from Admiral Yarnel.

"Where is everybody?" I asked. He looked at the silver eagle with the single chevron on my blues and said, "Nobody here but us shit-scratchin' chickens."

"Is'that so?" I replied. "I've come for a bale of blankets. Do you have a file on it somewhere?"

He pointed toward the laundry. "You must be talking about them slope-heads over there. They're washing and bailing them for shipment to Cavite."

"That's them! I'll pick one up and be on my way."

"In that rickshaw?" he exclaimed. "I was expecting a truck!" And then a suspicious look came into his eyes.

"Say," he said. "You got an invoice or something?"

I felt my positive approach crumbling. We were about the same age, and we looked at each other. He shifted his gaze to Li Teh. Her face was smiling and expectant, as she looked down from the rickshaw.

He turned back to me and asked, "She your girl?"

"Yeah," I replied.

"Nice looking chick. I got one in Tientsin." He hesitated a moment and then continued. "I'll tell you what. If yer girl can talk them slope heads out of a bale'a blankets, I'll just go into the booth here and write up in my log when you all come out, okay?"

I turned commandingly to the rickshaw coolie. "Chop Chop! Mush!" As though I were driving a team of Alaskan sled dogs.

Li Teh didn't have any trouble with the laundry people. The magic word was Doctor Graham and two dozen freshly laundered wool blankets were ours. We wrapped the bale in a rice straw mat and lashed it to the back of the rickshaw. It stuck out like the bustle on a 19th century woman's skirt, and Chop Chop had to shift his weight all the way forward. But that was okay. All three of us were happy.

And I had saved face.

"Number Three Beach," I ordered. "Let's go swimming!"

NUMBER THREE BEACH

▌0930 HRS.

As we pulled up the hill, behind the old German fort, Chop Chop was walking now. I leaped out of the rickshaw and pulled off my jumper.

"Li Teh, Tell him to get in."

The old rickshaw coolie protested but finally we talked him into the cushions while I placed myself between the shaves and walked up the hill.

He kept shaking his head, mumbling in Chinese, "Crazy Americans , they must eat loco weed."

From the top of the hill we looked down on Number Three Beach. In midsummer it was the place to go, with a composite mixture of people, cabanas and vendors. It was almost empty now. The early winter sun cast an orange glow upon the white sand but down beneath the hill a small cove looked warm and inviting and off the granite promontory nearby, white crested waves were breaking. They marched in even rows around the point, their emerald green shoulders glistened in the sunlight as they swept in an even predictable arc into the cove. It was a perfect body surfing wave.[15]

---

15 Tsingdao (Qingdao) is one of the few cities in northern China where surfing is possible. Qingdao's south-facing beaches provide the area the best wave and wind orientation for surfable swells. The best surfing waves form and break during the typhoon season (June–October).

[57]

I gave a hoot and ran down the hill. It was almost a disaster.

The rickshaw suddenly became the fastest rig in China, taking all of my weight and strength to overcome my lack of technical skill.

Chop Chop gripped the tongue and shouting instructions while Li Teh laughed hysterically.

We bought a shellaced duck and several quart bottles of dark German beer and spread a mat beside a cabana in the cove.

From this vantage point the waves were definitely larger, blue green walls with a nice even shoulder. They were the best waves I'd seen since leaving Hawaii.

While I slipped into my trunks and studied the waves, Li Teh unbuttoned her gown. Beneath it, she was wearing a one-piece bathing suit. The kind that was in fashion on the coasts of CA. It was a yellow satin material and with her smooth olive skin it was perfect.

As I stared, she held up her arms and pirouetted before me. "Do you like it?" she asked.

"Ravishing!" I said.

"Ravishing?" she inquired. "What do you mean by 'ravishing?'"

"Your suit, it's beautiful!"

"I copied it from a magazine," she replied.

And then I knew it had been Li Teh on the sewing machine last night.

She smoothed her sleek black hair, inquiring coyly, "Am I as pretty as American girls?"

"*Ding-Hau*" I exclaimed with sincerity. "You are prettier!"

"I think *Bu-Hau*," she laughed. "You're just like Chinese men. Talking nice just to make me happy."

"Do you swim Li Teh?" I asked.

"No," she said. "Do I have to?"

"Not if you don't want to. You can watch me bodysurf some of those waves out there."

A little frown wrinkled her forehead "What do you mean by bodysurf?"

"Ride them in on your stomach like a surfboard."

Her frown deepened "What's a surfboard?"

"You wait here, where it's warm," I replied, "and I'll show you."

> *Those waves were perfect vertical green walls. They seemed to hang while my body dropped in down the wave's face, and I could stay out on the shoulder with the hook hanging over head, the wave breaking and roaring behind.*

    I walked out to the end of the point and waiting for a lull I dove in. The water was cold, down in the middle fifties and I came up gasping and swam hard for the outer break to warm up.

    The waves were perfect vertical green walls. They seemed to hang while my body dropped in down the wave's face, and I could stay out on the shoulder with the hook hanging over head, the wave breaking and roaring behind.

    After the first ride I gave a lonely surfing cry and plunged back in. The nearest surfer at that time was 5000 miles away. For wave after wave I hacked that cold water, never realizing I was numb, and when I looked up, a crowd of Chinese were standing on the beach.

    I waited for a last good one and rode it in through the shore break until I could stand up. As I walked to the cabana with Li Teh we were surrounded.

    "Where is your engine?" they exclaimed.

    "How do you steer?"

    "Why do you do it?"

    Li Teh answered for me: "He says he does it because it's exciting, like playing *Mah Jong*. It's good exercise, like *Tung Fo*."[16]

    "AH!" they nodded knowingly, but still they looked at me as though I were some marine mammal that belonged in the sea.

    Chop Chop did the driving when we headed back.

16 Majong, and Tai Chi Chuan

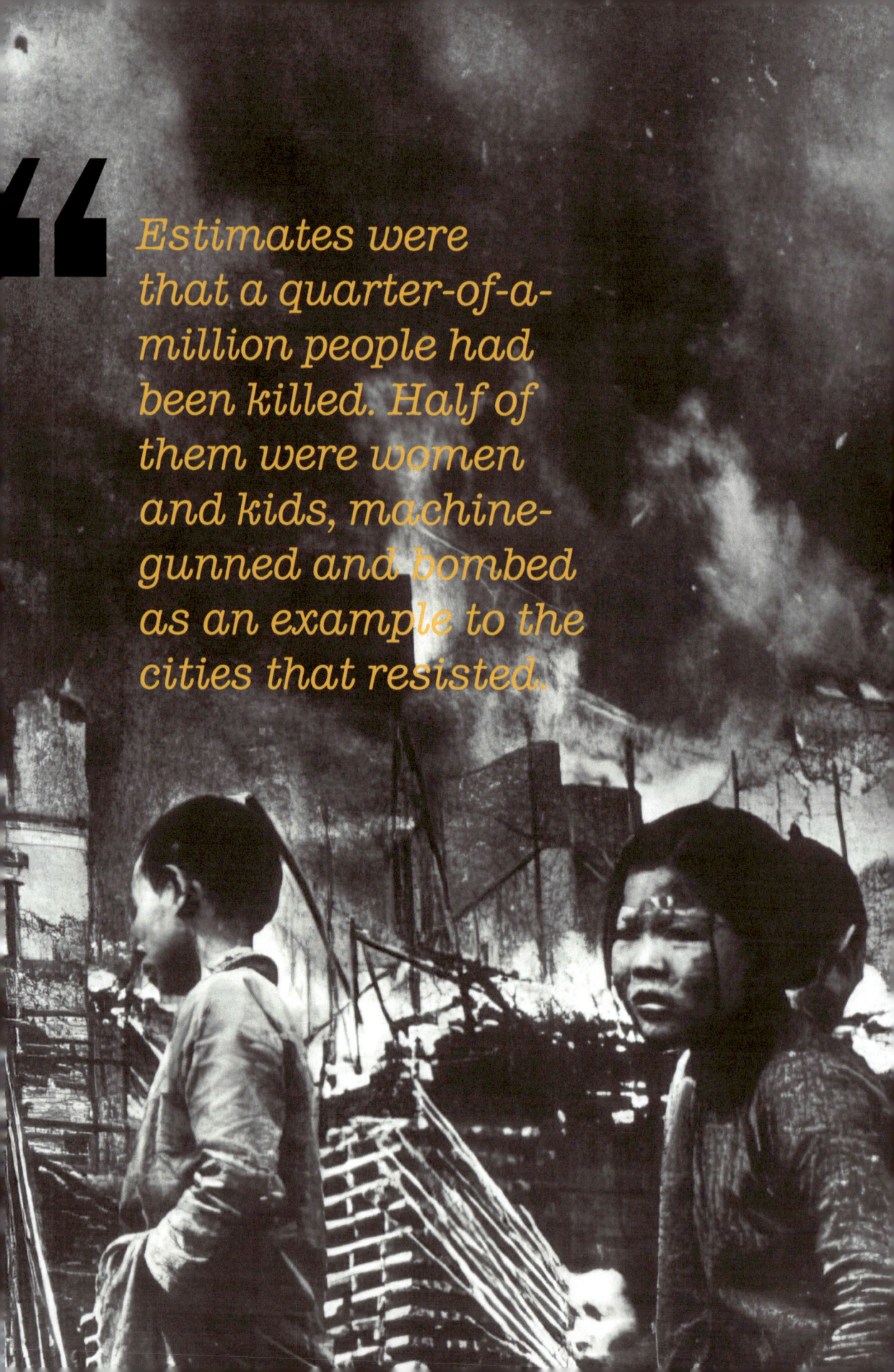

> "Estimates were that a quarter-of-a-million people had been killed. Half of them were women and kids, machine-gunned and bombed as an example to the cities that resisted.

■ 1745 HRS.

I gave Chop Chop some *mex*, a couple of Chinese dollars, to eat on while Li Teh and I went to the best restaurant in town. It was loud and noisy, but the food was good as usual. Won Tons, wheat cakes, sweet & sour pork, deep fried frogs' legs and shrimp with a crisp brown batter covering them.

I was living up to the image all Chinese had about Americans. We were millionaires no matter what our status and with the dollar exchange sixteen-to-one, I usually felt like one.

A group of Japanese soldiers eating in a corner didn't seem to slow things down much.

"How do you feel about them Li Teh?" I asked, nodding toward them.

"They are monkeys," she said. "We are peaceful people and they are doing terrible things to us. We are slow to anger, so now the Chinese are losing face. The time may come when all of us will be ready to die and the warlords of China will lead us. They can kill us fifty or a hundred to one and we will still win."[17]

I looked at her and was frightened. She could easily be a modern Joan of Arc, but she was too beautiful to die by the bullet. I thought of Shanghai and after that there had been Nanking. A quarter of a million people had been slaughtered if the reports were right. Half of them were women and kids, machine-gunned and bombed as an example to the cities that resisted.

China just didn't have the weapons or the organization to resist.[18]

---

[17] In June 1938, Jiang ordered the Yellow River's dikes and dam to be blown in a desperate attempt to slow the advance of the Japanese invasion. While this ploy worked, it was a disaster, causing flooding and killing between 500,000 to one million Chinese civilians. Up to ten million were left homeless and millions of acres of important farmland were destroyed.

[18] During the Japanese war against China, 54.4% of Japan's weapons and supplies were provided by Americans. 76% of Japanese's planes came from the US in 1938, and all lubricating oil, machine tools, special steel, high-test aircraft petrol came from the US, as did 59.7% of Japan's scrap iron and 60.5% of Japan's petrol in 1937.

1800 HRS.

We found Chop Chop, happy and standing by. He was a man of means with a few *mex* in his pocket.[19] He had dined on *chow-mein*, prepared by a passing noodle vendor. All fueled up and raring to go, he pulled us out of town toward the mission hospital and after we hit a level stretch I removed my jumper and shoes and took over.

I learned to adjust myself between the shafts until half my weight was off the ground. When I shifted close to the rickshaw it behaved like the fulcrum on a teetertotter and my feet left the ground.

Chop Chop giggled. It was a suppressed high note like a squeaking wheel.

"Mush!" he commanded. "Chop Chop," and then he giggled some more.

It didn't take long to work the excess energy off and I settled down. About dusk I pulled them into the compound.

Chop Chop, Mush!

---

[19] "Mex" was the acronym for "Military Exchange Currency," a form of script issued by the American military command.

MISSION: HOSPITAL COMPOUND

1900 HRS.

My forty-eight hour pass was almost over, and my ship would be getting underway in five more hours.

"Wonderful!" Doctor Graham exclaimed when he saw the blankets. "We'll have them made into pantaloons before you know it.

"How is Yu Fong doing?" I asked.

"He'll make it okay, but we have a problem. We had a visitor."

"Not the Japanese I hope."

"That's right," Graham said, "Captain Yataro Nagano the field commander for General Atigaki."

Graham shrugged his shoulders.

"He knows more about Yu Fong than we do. Our young farmer is a Communist, fighting for Mao. He was in a raiding party of some kind and

> *Our young farmer is a Communist, fighting for Mao. He was in a raiding party*

Nagano boxed them up in a village not far from here. He wanted to take Yu Fong out and shoot him right here in the compound. I refused, of course,."

"How did you stop him?"

"I radioed the American Consul in Peking who filed a protest to the Japanese government."

And then a worried look came into Grahams eyes.

"We're on the horns of a dilemma! Of course it's obvious, as Captain Nagano pointed out. If this mission is truly neutral we'll help no military people of any kind. And if we do, Iwasaki will be delighted to fill our wards with Japanese wounded."

Then he shrugged. "I guess temptation has been pretty well eliminated. They're going to post a guard outside of our property lines. They'll be screening all Chinese people coming in here."

2340 HRS.

It was almost midnight when I said goodbye.

It was clear and crisp with Orion and Sirius arcing through the sky. As Li Teh looked up at me, reflected starlight glistened in her eyes.

"I have read many American magazines," she softly said, "and all the time people in love are *kissing*." Her voice became shy. "What is this *kissing*? What do you do? How does it feel?"

So, I kissed her.

It was more of a caress, where I let our lips linger. Her arms hung loosely by her side so I slowly placed them around my neck and we kissed again. It was just a simple thing, more an exchange of affection with a little body contact.

But one thing for sure, the affection was real.

I wanted to protect her, take her with me. A fleeting thought crossed my mind which I quickly put down. *"How far would she let me go? No! Don't louse it up Brown.* Let sleeping dogs lie—don't turn the tiger loose."

"I like *kissing*," she said. "Will I see you again?"

"I hope so Li Teh. I go where the ship goes. There's a good chance we'll be back again."

"I'll wait for you," she replied. "I'll be here."

I jogged the first kilometer into Tsingtao I was so high. It had been a good weekend with some of the finest surf I'd seen in years. Riding those green walls after so long a time was as good as riding one of the cabaret girls out of Moppy Joes.

And then there was Li Teh-Chuan.

 RIGHT FULL-RUDDER!—
MEET HER!—STEADY

USS POPE DD225, UNDERWAY

0750 HRS.

The following morning our ship got underway with a mail pouch for the American Consul in Peking (Bejing). It was priority material under lock and key.

Captain Couble and the Chief Engineer were worrying about fuel and so we proceeded at a cautious fifteen knots. Rounding Shantung (Shandong)

Peninsula we passed Chefoo (Zhifu) and proceeded northwest—across the Gulf of Bo Hi, toward Taku Bar.[20]

As usual, we were "accompanied."

The bridge and masts of a Japanese cruiser was a hull-down presence on the horizon.

Because the Chinese had to eat, hundreds of fishing junks were scattered along the coastal waters. They came in all sizes from small, open craft, little more than big rowboats, to the large full-battened lug rigs with painted eyes on the bows and dragons carved into their high pooped sterns.

We wove through them dodging the gill nets and set lines. At night they marked their position with brush torches, ignited by the flaming sparks from a hibachi. The junks always waited until the last minute, hoping we would miss. Then suddenly there would be a ball of fire and a shower of sparks dead ahead. Then it would be "RIGHT FULL-RUDDER!—MEET HER!—STEADY—"

Then back to the course again.

---

20 Bo Hai was a shallow northwestern arm of the Yellow Sea, off the northern China's coast, enclosed on the northeast by the Liaodong Peninsula and to the south by the Shandong Peninsula, Liaodong to the northeast and to the south by Laizhou Bay which is generally considered part of the Bo Hai.

TIENTSIN

■ 1500 HRS.

We learned quickly that only thing that wasn't controlled by the Japanese, were the foreign legations and consuls.

Doctor Graham had become a minor international incident. The powers that be now felt the complete withdrawal of American forces left too much of a vacuum.

Our destroyer would now be designated as a "mobile-station ship" to provide at least some level of the previous status quo.

DOCKSIDE: TSINGTAO (QINGDAO)

1350 HRS.

We lay at the dock four days and for some of the crew it was the place they wanted to be. We all had our favorite ports, but when we received our orders I felt that Fengshui, the good luck spirits of wind and water, were really working for me. We were going to be stationed in Tsingtao for an indefinite period.

Calligraphy by Martin Sugarman

▎ 0800 HRS.

We moored to the dock beside the old, red brick warehouse and as we cleaned up the bridge we could see a part of Tsingtao (Qingdao).[21] I was becoming attached to it. It was a warm feeling flooding through my mind as though I were looking at my home town, back in Laguna Beach.

It was the surf, I thought, just around the point at Number Three Beach, but in the back of my mind was Li Teh-Chuan.

I had the duty that first night but the following day and night was all mine. Things had changed a little at the mission hospital.

A short, chunky, Japanese soldier stood alongside the road beside the main gate. He had a funny little rifle with a bayonet about six inches long. As I walked up and looked at him he stood his ground without moving a muscle. I had a sudden urge to stand him on his head and spin him like a top, or better still, jam the bayonet up his ass, clean to the stock. But he was only doing his duty, carrying out orders, even as I.

"How's business?" I asked, when I saw Doctor Graham.

"Slowing down a bit," he replied. "But you know how it is. People who are really sick or injured are going to come to me anyway. What is there to lose?"

"Yeah," I said. "When it gets down to it, life is pretty sweet no matter how tough it is."

When I asked about Li Teh, Graham smiled. "The rickshaw is all yours, take her into town if you wish. I have plenty of help."

He stepped back and looked me over. "They've been telling me about the things you did and my rickshaw boy wants me to call him Chop Chop now. Guess it reminds him of the old days when he pulled the buggies in Shanghai."

"He's one good Chinaman," I said.

"Yes," he replied, "most of them are."

Li Teh and I sat in the rickshaw hip to hip and leg to knee. It was a different feeling now. Without knowing it we were placing a mental wall around

---

[21] A major city in eastern Shantung (Shangdong) Province.

> *we were placing a mental wall around us, a protective barrier from the outside world. A euphoric state where evil spirits could not penetrate.*

us, a protective barrier from the outside world. A euphoric state where evil spirits could not penetrate.

The monsoon wind blowing off the Gobi Desert toward the sea had an icy edge to it now. Chop Chop wore a padded cotton jacket and pants. His horny feet were encased in leather sandals. Chuckling in his high squeaky voice he placed a lap robe across our knees and tucked it in.

We went to the top of the hill looking down on Number Three Beach. The waves were there alright—green and cold with the wind whipping the tops off them. We dropped back into town, went to a restaurant and tried a bar. Even though we were content, the day would become an endless round of the same thing. It would start to sag before we could get it off the ground.

"Li Teh," I said. "We need a snug harbor with winter coming on. I think I'll rent an apartment, a flat, something I can fix up and call my own."

She looked at me, at first gravely, then with a little smile.

"You don't have to live with me," I added. "I'm doing this for myself, a little hide-away from the ship."

"It isn't that," she said. "I was just thinking of a family I know. They have a store up on the hill and sometimes they rent the upper floor. You can even see your ship from there."

It was perfect. Fengshui had done it again.

It was well-built with casement windows looking over a part of the harbor. The opposing windows looked down on the busy street with its carts, stalls, vendors and city noises. Clean mats were on the floor. It even sported a *kang*, a boxed out Chinese bed normally seen in farmhouses.

Charcoal braziers could be placed beneath it keeping the occupant warm and snug for the night. Cushioned benches against the walls served as davenports.

A tiled counter served as a cooking area with water buckets pans and bowls. A potbellied, teakwood Buddha looked benignly toward the well-scrubbed table in the middle of the room.

The added touch was the private water closet with more crockery and bowls. The toilet consisted of a watertight box with handles and a lid. Nicknamed the "honey bucket" it could be carried downstairs and dumped in a barrel that would eventually be hauled away.

An added bonus was the landlord and his family. They were a happy lot, with kids who stopped running just long enough to stare at me with bright wondering eyes.

"How much?" I asked.

He looked at me apologetically. "One hundred and twenty eight yuan a month," he said.

"*Ai-eee!*" I exclaimed in mock gravity. "One hundred and twenty eight *mex?*"

I pulled out my wallet. It came to eight dollars American. I gave him two more dollars for a small room in back. Chop Chop now had a place to stay when he brought Li Teh into town.

We spent the afternoon going through the shops in Tsingtao, looking for the little things that Li Teh thought I needed. A teapot with a set of porcelain cups, a tapestry for the wall, a large quilted blanket as smooth as satin. She was a real shopper, haggling the storekeepers until I thought they were giving it to us for the pleasure of bargaining with her.

That night we ate in a little restaurant with only four tables, a snug little spot, and easy to keep warm. Li Teh was relaxed. She was happy. It made me feel good inside to see her smile. I hadn't mentioned going back to the mission and when I looked at her to do so she had been watching me. It was a studying look, intense, full of emotion. I placed my hand on hers. It was trembling slightly. Her kissable lips parted as though she were going to speak but there was no sound. I felt my pulse quickening almost keeping time with hers. What sweet excitement it was.

I paid the bill, gave Chop Chop a couple of mex, told him to get comfortable and walked her four blocks to my new home. We were in a daze, elbowing our way down the cobbled street. She knew what she was doing to me. She tried to hold her distance—still, she clung to me.

The landlord smiled encouragingly as we went up the stairs.

Standing in the middle of the room, Li TEH disrobed and pirouetted before me. She was exquisite with everything in the right places. Why is it that a woman can disrobe with such grace compared to a man's clumsy efforts?

I cherish the thought of Li Teh that night and all the nights thereafter. She was lovely, elegant and full of emotion.

ABOARD SHIP, PORTSIDE

■ 0845 HRS.

Our ship was rife with rumors the next day.

"We're going home to the States."

"Nope, we were going to join forces with the British in Singapore."

"Uh-uh. We're going to take Doctor Graham to Manila."

"Baloney. Y'a ain't gotta clue."

Scuttlebutt intact, an excited crew climbed through our ancient four-stacker, revitalizing her heart, bore-sighting her guns and polishing torpedoes.

On the bridge the steel splinter shields were in place.

We rove new steering cables to the steering engine room. Finally, there wasn't much else we could do except stand by.

And that's what we did in Tsingtao (Qingdao).

The good energy of Yin and Fengshui were still with me, despite Captain Yataro Nagano, who was doing his best to create an international incident. The American doctor had to go. His foreign ideology turned the heads of the Chinese.

And besides, Nagano thought, removing Graham would set a precedent. He would have a lever to get *all* of the irritating *"foreign elements"* out of China.

It was hard enough bringing these people to heel without this subversive foreign interference.

The wires hummed while we waited, a pawn in the ultimate decision.

Illustration by Kevin Short

MISSION HOSPITAL

▎1545 HRS.

I trotted out to the mission hospital as soon as I could get away.

Doctor Graham had more company now. Japanese troops were coming onto the premises, using the mission's water well and then they would sit around laughing and talking.

To add to his unwanted presence, Captain Nagano even invited himself to a surgical operation and congratulated Graham on his technical skill.

Still, the Chinese came, paying their bills with wheat, squash, peanut oil, anything they could spare. Doctor Graham was selling a commodity that was hard for Nagano to beat—love and affection and a doctor's skill.

"How are you cutting it Doc?" I inquired. "I wish I could bring a landing party up here."

"That would precipitate things," he replied. "Things will turn out okay. I've just received a wire from the Consul. They've submitted an official protest to the Japanese government"

He was a feisty rascal. The pressure didn't seem to bother him a bit. He almost seemed to thrive on it, calling the shots as he saw them while either Buddha or the Lord Jesus handled the scene.

And then there was the other side of him as he stepped back and looked at me. "There's enough raping and cutting going on around here," he said. "Can you take care of Li Teh for awhile?"

I looked into his steady blue eyes I suddenly knew there were no secrets in China. The Chinese were the biggest gossips in the world.

"I sure can Doc," I replied. "As long as I am able."

"How about Chop Chop?" he said. "He's devoted to her."

"I already have a spot for him," I replied. "He'll be warm and will eat well."

"Fine," he said. "She'll be less conspicuous until this blows over."

When we left, all of Li Teh's belongings were packed neatly in two bundles like a sailor on the move. And that night after we had settled down she slipped on an exotic Chinese gown. As I sat on the kang, smiling and admiring her, she kowtowed before me. Slowly she bent her knees, looking into my eyes. And then she bowed her head, placing her forehead upon the mat with her hands palm down, as though I were a Chinese prince and she was my number ten concubine. She crawled to my shoes began to unlace them.

"Li Teh!" I rasped. "You don't have to do this. I'm an American!"

"Hush!" she said. "Tonight I want to be your Chinese lover."

"Hush!" she said. "Tonight I want to be your Chinese lover."

She had the last word. I was caught in her spell.

She bathed my feet in warm, perfumed, water. She helped me into a Chinese robe as though I were a sleepwalker and for once I felt I'd done it well, as though I were in another world where all things were perfect.

She bathed my face and arms with a fresh bowl of water and all of our problems seemed to fade into a distant fog. It was just Li Teh and me. Nothing else mattered.

The following week was a continuation for us. I gave Chop Chop some wax and a can of bright-work brass polish for the rickshaw. He went a step better, hammering out a pair of copper dragons that were riveted to each side. When we came down the street at night with our carbide lamps glowing, the shop light glittering on the dragons and Chop Chop honking his horn, I felt like an Arab potentate sitting on a billion barrels of crude. We had the best rickshaw in town.

We went to Chinese theaters, performed by live people with chalked faces. Their costumes were colorful, and their voices had a tremendous range, from garrulous guttural to high C. Most of the plays were heroic with sad endings. It was easy to spot the villain. He was the guy with the turned down painted mouth, his face shaded into evil, learing lines. As Li Teh sobbed, I quietly held her hand as we compared him with Captain Nagano.

The Japanese field commander had removed his men from the mission compound, but now he was fomenting trouble in Tsingtao. All non-Chinese military personnel had strict orders to avoid any type of aggressive contact with Japanese troops while on liberty. It was "Peace, brother, at any price."

Captain Couble had said the first man to raise a fist would get a summary court martial, liberty would be curtailed, and he had always been a man of his word. A U. S. sailor can make a lot of sacrifices before giving up his shore leave.

We had done a fairly good job avoiding the Japanese. With the exception of our armed shore patrol, we were still going ashore unarmed as though we were in the States. It was the same for the Japanese who had their armed military police.

> *They still came in unarmed but when they saw any of our people walking down the cobbled streets the Japanese locked arms and came toward us three abreast, chattering as though they hadn't a care in the world.*

But those troops who were out on the town up until now were more or less the same as we were; "Troops, out on the town."

Only now they were using a different tactic.

They still came in unarmed but when they saw any of our people walking down the cobbled streets the Japanese locked arms and came toward us three abreast, chattering as though they hadn't a care in the world.

Obviously, they had their orders.

But for us? It was the Mahatma Ghandi strategy:

"Sailor, flatten against a wall or duck into a stall."

When this failed to arouse any confrontation, they would go into a cabaret and push the girls around. Nagano's Bully Boys were some of the nicer terms we called them.

They shouldn't have done it.

When it came to a brouha, a good old rough'n'tumble free-for-all, our ships company was hard to beat. Before the Japanese moved in we used to go ashore and fight each other for the hell of it. On occasion we'd set up a boxing ring with another ship and we'd have a smoker, using gloves and three-minute rounds, all legitimate like.

On goodwill tours we went ashore like stray tomcats—stalking our new territory, tangling with the "Laddies from Hell," the kilted Scotch highlanders stationed in Shanghai.

We'd bust a few noses and receive a few love taps, getting acquainted with the French Foreign Legion in Saigon. We used to bust a few noses and receive a few love taps, getting acquainted with the French Foreign Legion in Saigon. We hadn't had our pleasure with the Japanese and we were losing face with the local Chinese.

I guess Tex Gibson, the gunners mate, was lead off man. It was on my duty night and we were sitting around evening mess.

"Screw it!" he said. "I can't take much more of this."

He went ashore with some of his buddies and sauntered through town. When Nagano's Bully boys came down the street Tex walked into them head on. When he came back to the ship he said, "Hell! A wet noodle can fight better than them."

"Jiu-jitsu? I never saw any of it."

Tex was restricted to the ship and the whole thing became a sort of a hockey game. You went onto the field and took your shots, and if you were caught? You went to the penalty box.

I didn't mess around with the Japanese. I had other things on my mind. If they came at me in a direct confrontation that couldn't be avoided without eating dirt I made it as quick as I could.

A kick to the knee and when they were down a shot to the ribs or the head. They didn't know anything about boxing. I had a six inch reach on them and I could usually nail them with a few short hooks.

If there were not more than three I could be out of there in less than a minute, slipping through a shop into a back alley while the Chinese patted me on the back and whispered, "*Ding Hau! Tai-Pan, Ding Hau!*"

It wasn't long before Li Teh began checking my knuckles when I came in. The shopkeepers latched onto me as though I were a neighborhood folk hero and when she went grocery shopping they told her about my episodes. To avoid the penalty box, I began sneaking through back alleys to the flat.

The alleys of a Chinese city have to be experienced rather than told. With little or no sewage system the honey buckets and accumulating barrels of human excreta held there give off a stringent odor that will floor a man the first time around. It would turn paint yellow, raise the grain of untreated wood, kill living plants, and if it hadn't been for Li Teh I would have used the main drag.

For several weeks gentle spirits looked over us. And then one night, Chop Chop brought the rickshaw around to the front door. We were going out and while he lit the running lights, Li Teh was dressing.

God! She was beautiful. She had class, all the way.

As I watched her dress I heard a splintering of wood and a shriek from Chop Chop. It was a shriek of anger and dismay. "Ai-eee!" it was almost a battle cry.

I dashed down the stairs and there was Chop Chop, lying in the gutter, it just took a second to take it all in. Two Japanese military police were standing by the rickshaw with foolish smiles on their faces. They had jumped on the long tapering shaves that Chop Chop had felt between his hands so long and had shattered them. The rickshaw tilted weirdly, with its stern in the air, the carbide lamps glowing in his face.

> *they stuck him with their bayonets. A gusher of blood squirted from his neck, a great red blob spreading across his chest.*

Undoubtedly, he had protested. Perhaps he had charged them with his bare hands and they stuck him with their bayonets. A gusher of blood squirted from his neck, a great red blob spreading across his chest. There was no time to lose and I swept him up into my arms and carried him to the room. I struggled to pinch off the artery while Li Teh cradled his head in her lap. I didn't have a chance. Chop Chop was only a rickshaw coolie but he carried six quarts of blood, just like the rest of us. It all ended up covering our hands and our clothes.

I felt a volcano building within me, and anger I had never experienced before. I wanted to scream, smash chairs, kill, anything to release me from it's grip.

When I saw Li Teh sitting there, the tears streaming down onto Chop Chop's still face, I declared war on the Japanese three years before Pearl Harbor.

I put myself together while we cleaned up the place. Chop Chop stayed with us that night. Reporting his death would only have caused Dr. Graham and the Mission suspicion, and who knows what kind of trouble.

So, we covered his body, and Li Teh Chuan prayed. Looking at him lying there, I wished I had spoken with him more, made a better place for him.

The next day, with help from the landlord and a professional funeral parlor we gave Chop Chop a number one funeral.

The procession led off with a band of drums, gongs and cymbals. They were followed by Chop Chop in his wooden casket. With six Chinese pall bearers, he traveled in class.

Firecrackers were exploding, paper dragons floated over his head and around him were symbols of the things he would need, a beautiful paper house, a rickshaw with someone else pulling it, food, paper prayers of wisdom to satisfy the inner-man. They were all burned with full belief that somehow, they would materialize for him in another dimension.

Chop Chop was on his way.

But the cruel faces of the two Japanese military police were etched in my mind like a photograph.

I considered ways to send them to their ancestors.

Still, if anything went wrong it would be the penalty box or another international incident, adding to the accumulating score.

And worse still, our ship's entire crew would be restricted. What I was seeking was cold revenge.

I finally bought a small leather sap that almost fit in the palm of my hand. A soft ball of lead was stitched into the end of it, and it hung in my sleeve without being seen.

▎0330 HRS.

It took a couple of liberties over the next weeks in order to find them, and when I did I stalked them from the alleys, checking their beat and patrol pattern.

They hadn't changed any, shoving Chinese aside with their rifles as they came through.

I staked-out in a dark alley, and when they passed I jumped out in a blur of speed, and sandbagged them, smashing down brutally—swinging across their heads with the full weight of the sap.

Their rifles clattered into the street as they lost consciousness.

It was time to settle Chop Chop's score.

Kicking them into the alley, I worked them over.

Bone crunched beneath the lead-weighted sap. When I felt their teeth snap it made me feel good, releasing the nightmare tensions within me.

Bloody, beaten, unconscious probably with concussions and broken facial bones, crushed nasal cartilage, missing teeth, eye-sockets fractured, they lay urinating in their pants.

As I looked down on their unconscious bodies the steam left me.

With the release of my anger came clarity.

This was over. It was cold-blooded assault and it was a one-time deal. I didn't have the guts for more of it.

In exasperation I looked up to see the expectant faces gathering in the entrance to the alley. Picking up a honey bucket of excrement I hoarsely growled in English. "What's the matter with you people? You don't have to take crap!"

I poured the sickening mess of human excrement over the two unconscious men. I hadn't killed them but the probable infection resulting from Chinese fecal matter probably would.

▌05 NOVEMBER, 1938

It was a fine opportunity for Captain Nagano to create a propaganda incident.

His first sentence could have read: *"On the night of the fourth day of the eleventh moon American sailors attacked Japanese military police."*

But no such message was forthcoming. It was a matter of face, again. With Japan's mythic tradition of skilled and invincibly victorious Samurai warriors, General Ataki's just couldn't acknowledge the maiming of two armed military policemen by an unarmed American.

In the long run I knew I had made a mistake, playing into Nagano's hands. It was Yu Fong who gave me the message. He was no injured farmer now. Alert and shrewd he came to our room.

"Captain Nagano has complained to the Tsingtao, Chinese police," he said. "Protesting the lack of cooperation. He's going to declare martial law again. He will get to the Chinese through the Americans."

For a while I thought Yu Fong belonged to a secret organization, maybe a member of the Green Dragon Society, or a hatchet-man for the Tongs, but as he explained, those organizations had submerged, forming underground cells inside the Japanese lines. They worked for the Kumintang under Chiang Kai-shek, or Mao and the communists. Both parties had declared a truce while they fought the common enemy.

It didn't take long for Nagano to make his move. Within six hours the various Chinese puppets, department heads appointed by the Japanese were removed, sand-bag machine gun nests sprang up at strategic street corners.

And then he concentrated on Doctor Graham.

His dispatches to the Consul and Captain Couble described the necessity of removing the good Doctor.

"We can no longer be responsible," he explained. "We cannot guarantee his safety."

Obviously, Nagano, or anyone else for that matter had anticipated Doctor Graham's move. He filed Chinese citizenship papers with the Nationalist Kumintang, thus releasing the American government from all responsibility for his safety.

His letter to all concerned parties stated that although he loved America he would spend the remainder of his life with the Chinese people with whom he had worked so long.

▌0800 HRS.

Li Teh and I started putting things together again.

She burned incense before the little Buddha in our room and she whispered Christian prayers. She told me more about Doctor Graham, explaining how he lived from day to day, making every one of them count, no matter what the circumstances.

And at the end of each day he thanked his maker for the privilege.

"I have seen him with my own eyes!" she exclaimed. "Kow Towing in front of his bed." She held her hand over her heart. "Jesus is inside of him. Are not all Americans that way?"

"Not quite, Li Teh," I answered. "Some of us like to scull our own sampan, and not leave things to fate."

"What do you mean fate?" she persisted. "Doctor Graham has never said anything about fate and he is happy inside."

Li Teh was the same way. She hung in there, leaving nothing to chance, acknowledging two deities even occasionally placing a slice of tangerine or some morsel of food beside the Buddha for Chop Chop's spirit.

2000 HRS.

That night a nimbus veil of clouds moved in and with it came the rain along with our ship's orders. The shore patrol passed the word.

Going through the bars and cabarets, the butts of their .45s sticking through the slot in their black rain coats, they found me with Li Teh. We were eating dinner with the landlord and his family.

"We're shoving off at midnight!" Tex Gibson said. "We're heading to Manila."

My heart tightened as though it were in a vise. "Maybe," I replied.

"What do you mean maybe?" Tex exclaimed. "If you're not aboard by twenty-three hundred I'll be back here with the whole gang!"

The remainder of the dinner was like the last supper. I felt like Judas about to betray these fine people. I could hardly look into their eyes.

Tex was right of course. Even if I did jump ship, the word gets around in China. I'd be on the run, hiding in cellars, hiding from my own people, not to mention the Japanese. It just wasn't my way of doing business. And Li Teh, I knew, wouldn't either accept or approve.

In our room we made plans.

"I'll send for you Li Teh," I said. "And until I do you'll receive a check every month. It'll cover all of your needs. I'll see about your passport in Manila. And if something should go wrong I'll come back and get you. My enlistment expires in two years. We can wait that long if we have to."

The words poured out of me. I couldn't stop them.

And Li Teh held me closer, reaching up to me with her sweet lips. We were young, the ecstasy of our love, the bitter-sweet anguish of parting transported us to heights we had never experienced before. Our bodies seemed to be floating through space, her heart beating against mine like the wings of a startled dove.

As time ran out I said, "I have to go Li Teh."

"I'll walk down to the dock with you."

"No, best not," I replied. "It's raining. You're warm and safe here."

She shrugged her shoulders as though nothing mattered after I left. "Just a little while longer," she pleaded. "I'll be good."

We slipped down the stairs through the back alleys to the warehouse where I had first met Doctor Graham. We looked at the destroyer getting up a head of steam. The blowers were humming while black smoke quietly oozed from her stacks. Rain glistened on the bright steel of her quarterdeck. A dim light glowed in the wheelhouse.

She bowed before me in the rain. "Good wind and water," she said in Chinese. Her voice was soft and firm, as though I were the master of a great Chinese trading junk on a two-week cruise down to Ningpo. "May your voyage be prosperous."

"Thank you," I replied. Stupid words! There was nothing in my vocabulary to express my feelings. "Remember what I said Li Teh. Keep our room. I left money under the Buddha. Enough to last you for three months and I'll send more."

Then her defenses crumbled. That's the way I remember Li Teh-Chuan standing in the shadows against the brick walled warehouse the tears and the rain washing down her cheeks. I didn't know a man could hurt so much and still live.

PART II

WAVES OF WAR

"CONGRATS, BROWN."

By mid-1938, and then as 1939 unfolded and political tensions rose, history was happening fast—all over the world.

Adolf Hitler and Josef Stalin overran Poland in September. Now Hitler was looking toward France and England, the Wehrmacht, his Blitzkrieg army, Panzer tanks and Stuka dive-bombers wheeling, while his new U-boat fleet was gathering strength beneath the North Sea.

It didn't take a student of military history to divine his plan. How many other countries had allowed their leaders to develop grandiose aims and military power and never used it? Not one.

From Genghis Khan, to Alexander, to Napoleon. While diplomats negotiated, nations began arming, testing their mechanisms of defense. Military conscription, advanced tactical vehicles, new ships, guns, aircraft. The development of a military juggernaut takes human sweat—and taxes. It's like rolling up a snowball. Once it's built it has to start rolling or it will melt and collapse of its own weight. It has to move, feeding off its host or it will become an anchor around the neck of the nation that built it.

While diplomats negotiated, smart countries began setting up defense mechanisms. The draft, mobile vehicles, new ships, guns. People were like worker ants, miniscule in their individual efforts, yet necessary for the survival of the colony.

Old Uncle Sam was one of the smart ones. He started out by building a few new destroyers. The Maddox Class they were called, two stack 1800 ton jobs with 45,000 horsepower, long and lean with low topside weight.

Of course none of us knew much about this until we pulled into Manila. China had insulated us from the outer world. The crew acclimated to their new environment like syphoning oysters in a new bed.

" *The ship slipped smoothly out of San Diego Harbor*

Men were going up in rate all over the Asiatic Fleet.[22] Eight or ten were pulled off of each ship and transferred to the States.

One day the division officer handed me a slip of paper. "Congratulations Brown," he said. You're a First Class Quartermaster now."[23]

He handed me another slip. "Here are your orders you lucky dog. You're headed for the Bremerton Navy Yard. One of those new eighteen hundred tonners, the Charles F. Hughes."

My heart sank. Tsingtao would be farther astern over the horizon.

---

22 "Going up in rate..." A U.S. Navy *rating* is defined as an Occupation, consisting of specific skills and abilities. Each rating has its own *specialty* badge which is worn on the left sleeve by all qualified men and women in that field. Reference: U.S. Navy Enlisted Rating Structure - Bluejacket

23 U.S. Navy First-class Quartermasters serve with officers of the deck and Navigator's Assistant. They also serve as Helmsman and perform ship control, supervise the ship's signal force, navigation, and bridge watch duties. Quartermasters can also see assignments to sea duty as petty officers on seagoing tugboats.

ABOARD CHARLES F. HUGHES

■ 0840 HRS.

    Things started looking up again after I reported aboard. There's plenty to keep a man busy with new construction.
    Being a newly commissioned ship meant we held our builder's trials in the Straits of Juan De Fucha off the coast of Washington. With the break-in trials concluded we proceeded south, to San Diego.

upon completion of shakedown, the ship slipped smoothly out of San Diego

Obviously we were going to head west.

The ship had a full set of charts for the Pacific including China. We had only token sets of charts for any other place in the world.

One evening about 2000 hours we anchored at San Clemente Island.

The light on China Point was flashing through the darkness a mile away. It was appropriately named. I broke out a harbor chart of Tsingtao and plotted our room, four blocks North and East up the hill. It would be high noon in Tsingtao, tomorrow. They were across the dateline.

The sun would be shinning directly down on Li Teh, and I was hoping she would be having lunch with Doctor Graham. He was a man for all seasons, as the saying goes. In the long run, how could anyone harm a man whose heart encompassed the world? He was universal with a simple, direct, philosophy. *"Help thy neighbor regardless of race or creed."* I could still hear him saying, "I've seen the insides of about every nationality in the world and we are all the same."

Upon completion of shakedown the ship slipped smoothly out of San Diego Harbor. Between a round of bearings the navigator turned to me and smiled, "As soon as we clear the sea buoy, break out portfolio thirty-nine."

PORTFOLIO 39

"Portfolio thirty-nine? Those are the General charts for the East Coast!"

"I know," he said. "We're headed for Old Gitmo Bay, Cuba."

Christ!

One hundred and eighty degrees out of phase. The shortest way to China would be through the center of the earth.

> *From that time on, the Charles F. Hughes never stopped moving. From airplane guard and escort duty for the East Coast carriers to simulated depth-charge attacks against our own submarines.*

From that time on, the *Charles F. Hughes* never stopped moving. From airplane guard and escort duty for the East Coast carriers to simulated depth-charge attacks against our own submarines. At night gunnery exercise, daytime antiaircraft and torpedo practice drills.

We saw a few ports but never for longer than a few days. Charleston, Norfolk, New York, Newport, and once in a while we would pull into the Boston Navy Yard for experimental equipment. The Navy was entering the electronic age.

Our crew was honed down, by consistent drill and practice, to a smooth weapon. Defense or offense—it didn't make much difference.

300 MILES SOUTH OF ICELAND

■ 07 DECEMBER, 1940

With Congress passing, and President Roosevelt's signing of the Lend Lease Act in 1940, one year before Pearl Harbor, the shooting war began for us.[24]

We formed a division with the destroyers, *Lansdale*, *Niblack*, and *Monson* and commenced escorting British convoys across the North Atlantic until we knew the way like the backs of our hands.

Cold; the smell of an unseen iceberg in a snow flurry; icing decks with the Northern Lights rippling across the sky like green, purple curtains, we were living for weeks and months under combat conditions in our ship's sealed compartments with dim red battle lanterns providing our only light.

The flaming roar of an exploding oil tanker, when a Nazi submarine's torpedo connected, the *kurumpf* of the depth-charges, the eerie sound of the pinging sonar when our electronic detection equipment located a submarine, deep below us.

When Pearl Harbor was hit we were 300 miles south of Iceland. Torpedo Junction we used to call it. Out of 70 odd ships we had lost more than half, torpedoed and scattered across the breaking seas. I was on watch on the darkened bridge when word came up from the radio shack.

"Jesus, God! The Pacific Fleet sunk?"

For awhile we couldn't believe it.

But war settles in fast.

---

[24] During March of 1941 the U.S. Congress passed the Lend-Lease Act enabling England to obtain merchant ships, warships, munitions, and other materiel from the U.S,. to assist with the war effort.

BAMBOO TELEGRAPH

I still had a pipeline to China of sorts.

The bamboo telegraph was still working. Chief Quartermaster Shockley, who had stayed on in China was a Lieutenant now. He was navigator on the *Lansdale*. When our ships fueled and provisioned in Iceland we would hash over the old days.

"You remember that pretty little Chinese nurse that worked at the Mission Hospital? What was her name?"

"Li Teh," I said.

"Yeah, that's right, you were going with her. Well, when I left she was still with that Doctor, Graham was his name?"

When we pulled into Argentia, Newfoundland I met a pilot who had been flying the Hump—hauling supplies over the Himalayas into Chungking (Chongqing). He was on anti-submarine patrol now pushing PBY's for the U.S. Navy.[25]

"Who? Doctor Graham?" he said when I asked. "Oh yes, I've heard of him. They oughta build a monument to that guy as an example to all politicians."

And then there were the letters I received now and then, care of Fleet Post Office New York. They were in English in beautiful square block print, carefully done with a Chinese writing brush. As of the last letter, Li Teh was receiving my thirty-five dollar money orders. Perhaps things were okay after all.

I worried too much.

But Pearl Harbor was a real incentive.

Ships were coming off the ways faster than trained men could man them. Ratings were fast, with more qualified personnel receiving commissions. Now was the time to make it. We kept plodding along between Newfoundland and Iceland, living charmed lives while our shipmates on the *Reuben James* and the *Niblack* were torpedoed.

---

[25] The Consolidated Aircraft's PBY Catalina flying boat is an amphibious aircraft produced in the 1930s and 1940s—widely used during world War II in every branch of the United States Armed Forces PBYs served until the 1980s. In 2014, the aircraft i s still flying as a firefighting watertanker.

Fresh destroyers started coming North with every convoy. Their virginal crews looked at us, at our ship's sea torn gunwales, rust streaked superstructure and dented sides. Our crew was in the same condition. After eighteen months on the North Atlantic, it ground a man down.

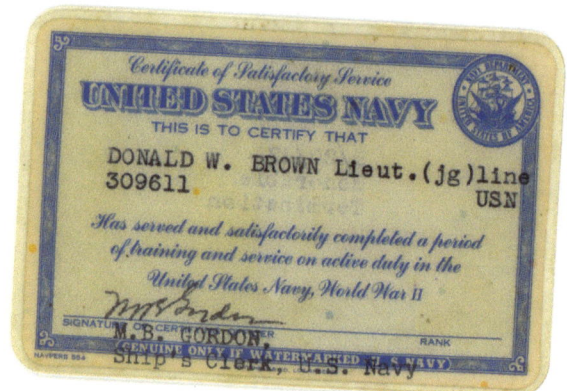

It was a lifetime.

Somehow the proper people got the word. Admiral King wasn't so bad after all. The *Charles F. Hughes* received her orders to proceed independently to New York. What a beautiful trip that was.

Sliding along through the wind and sleet then over the Grand Banks into the fog. We had no one to worry about but ourselves, and at the end of it was New York.

When we arrived the Bureau of Personnel was waiting for us, scattering the crew to the far ends of the world, mostly to tropical duty where we could get the ice out of our hair.

Some of us went to ships that would eventually go to the Mediterranean. Some went to the Caribbean and others to Diego.

I went south to the Dutch West Indies as Officer In Charge of a tug named *Squanto YT194*.[26]

We worked with Dutch civilian skippers that had run their North Sea tugs from Holland when the Nazis invaded the low countries.

Those broad assed, blue eyed sailors were *real* seamen.

Then there was a civilian skipper who operated a salvage vessel for Merrit, Chapman and Scott. They were assigned to the South Atlantic, under government contract.

---

26 *Squanto* (YT-194) was built in 1942 by Ira S. Bushy & Son, at Brooklyn, N.Y. She was placed in service early in 1943 in the 10th Naval District, covering the area around the Virgin Islands. *Squanto* was redesignated a large harbor tug, YTB-194, on 15 May 1944. She was reclassified a medium harbor tug, YTM-194, in February 1962. **Naval History and Heritage Command.**

Between towing and salvage jobs our tugs would rendezvous in places like Trinidad, Curacao or Aruba, and I would pick their brains over a bottle of Bols gin.

It was worth 120 units in any college, and at the end of it the Navy gave me a commission. I was now Lieutenant Junior Grade, Brown, USN.

They rounded me off with a fast course in administration. Amazing what those military service schools could do under pressure. They stuffed your brain as though they were loading an old muzzle loader, ramming the charges home, then tamping them down with a bit of wading.

TUG-OF-WAR

■ 1942

The Squanto was a deep full-bodied vessel and shiny as a new silver dollar, a step-deck rig with a high, flared bow similar to the North Sea tugs. She looked restless and powerful.[27]

I can still feel the throb of her diesel electric main drives as though she were a train, pulling a mile-long string of freight cars through the Rocky Mountains.

The barge we would be towing was just about as interesting. A rather new concept, a floating hotel two stories high. She had a crew of cooks and bakers, a large mess hall, Master at Arms Force and a Warrant Bosun in charge to keep things running.[28] In the Pacific she would become a portable barracks, replenishing the needs of the fleet.

It was a long grind into the South Pacific. Ninety-two days to Espiritu Santo in the New Hebrides, with fueling stops at Panama, Diego, Hawaii and Samoa. As we strained against our hawser, the great sea battles Guadalcanal, Savo Island and Cape Esperance were being fought.

It didn't take long to find out what it was all about. When we reached Santo, little squibs of information were received on general news broadcasts from Hawaii and Australia.

We patched up a sinking destroyer then moved NW to Tulagi, across from Guadalcanal.[29]

---

[27] Sea-going U.S. Navy tugs were capable of speeds of up to 16-knots, and were armed.

[28] Warrant Officers are the highest Enlisterd-level U.S. Navy rating.

[29] Guadalcanal Island was a pivotal battle in the quest for victory against Japan in the South Pacific in 1943.

A sunken Japanese destroyer had been spotted. Her code books would be invaluable if they were still aboard. We broke out the hard hat suits enroute, and moored over the wreck.

She lay at a 45-degree angle in 120 feet of water. The underwater lights and cutting torches were rigged and the bosun went down. He checked the open bridge then dropped to the radio shack.

"I think we've got em, Captain," he said over the phones. "They're all locked up neat like, in the safe. Send me a wire and we'll winch it out of here."

As the hooking wire went down I visualized his surroundings, the glow of the torch cutting the brackets.

And then the bosun came on the radio again.

"Say Skipper! I think I've got a present for you. The radio gang are still here all gassed up in their life jackets and floating against the overhead. Must be a pretty fresh wreck!"

A peculiar sensation swept over me. These were the first Japanese I had met in over two years, even if dead.

China was getting closer.

An impossible dream under present conditions. Still, we were back in the same ocean.

■ AUGUST 08-09, 1942

The months moved on while men and ships poured into the South Pacific. Admirals Halsey and Mitcher gave everyone a little fighting room after the second battle of Savo Island. It's funny how ship's crews think of these old sea dogs.

You begin thinking about them as though they were personal friends. Men with courage who lay it all on the line and it's very different with ships. They're right out there at the end of the spearhead with their men.

If I met Admiral Mitcher I'd probably greet him like a tug boater and say "Hi Mitcher, you old rascal. How's business?"

And I'd probably be reamed out good for it.

After the amphibious landing forces were formed into their various flotillas and squadrons, then came the landings, Bougainville, Tarawa, and Leyte.

It was a three-pronged effort that men on their own individual ships never had knowledge of. It was a matter of keeping afloat, keep the engines running and doing your job.

Amphibious landings became standard procedure. Several months of build-up then the adrenalin-pumping push farther north, Eniwetok, Guam, Manila, and Okinawa.

We were north of the Equator now, working with the spearhead, pulling hung-up landing craft off the reefs. Those that couldn't make it on their own we towed to the nearest repair facility. Many of them were floating cities, protected from the sea by a coral atoll.

Upon arriving in Pago Pago we looked at our tow distrustfully.[30] The IX 16 had a big problem.

It was the *materials* she was made of. She was a Liberty Ship whose hull was made of concrete. Her sides appeared to be 12-in. thick and she sat at her moorings like an immovable fort daring us to move her. Liberty Class ships were designed to fill a void in merchant marine ships as the United States rapidly ramped up its war effort. They were given the designation "EC2-S-C1," the EC standing for emergency cargo. With no propulsion system of her own, her interior spaces were filled with rows of walk-in chill boxes loaded with beef from Texas, Idaho potatoes, frozen strawberries, peaches and plums from the San Joaquin Valley.

Her diesel-powered refrigeration system poured a steady roar of sound through exhaust ports let into her sides. She carried her own crew of supply officers, diesel enginemen and refrigeration experts. They looked us over while we shackled into her, as people study their pilot when they board an airliner.

The IX 16 towed well-enough in calm water with a our sustained speed of 6 knots. When the seas started picking up she became a real bitch, sheering out to port or starboard at a 30-degree angle, cutting our speed to four.

She was tough on our towing gear and our diesels were getting tired. For two weeks we nursed her along, out of one rain squall and into another and sometimes our speed was down to three knots.

---

30 The ships were designed to minimize labor and material costs; this was done in part by replacing many rivets with welds. This was a new technique, so workers were inexperienced and engineers had little data to go on. Many early Liberty ships were affected by deck and hull cracks and indeed several were lost. About 1,200 ships suffered from cracks during the war (about 30% of all Liberty-class ships), and three were lost when the ship suddenly split in two. Though the work force was largely untrained in the method of welding ships together, it was not worker error that caused these failures. Rather, the failures were caused by a design oversight.

AUGUST, 1945

▌1730 HRS.

One late afternoon in August 1945, we shortened our tow off Eniwetok Atoll.[31]

We had been out of touch for a long time. Across the reefs inside the lagoon a massive fleet lay at anchor.

A string of aircraft carriers lay in a long line as far as the eye could see. Killers Row, we called it.

The atoll was jammed with 55,000-ton battleships, heavy cruisers, light cruisers, destroyers, tankers, supply ships, tenders, landing craft, and barges. All of them manned by thousands of men.

The bridge gang tried to total them up and they always lost the count at over 200. There was no way to see them all.

We found a small open spot for the IX 16 and anchored her.

After our main engines were secured, no one could leave the decks. We stared in awe at the mighty fleet America had produced and this was only a small part of it. That night we sat around the bridge feeling the security of the calm lagoon, almost refusing to believe it. But they were there alright. Their blinker lights were flashing, inquiring about our steaks, and Idaho potatoes.

No one was paying much attention to blackouts anymore. Stars were shining and we were feeling good. With power like this the war should soon be over.

And then an odd thing happened.

---

[31] Enewetak or Eniwetok is one island of a large coral atoll formed of 40 islands in the Western Pacific Ocean. As a legislative district of the Ralik Chain of the Marshall Islands, with a land area total less than 2.26 square miles, it is no higher than 16.4 feet. Surrounding an enormous, deep central lagoon, 50 miles in circumference, it is the second-westernmost atoll of the Ralik Chain 190 miles to the west of Bikini Atoll.

> *"This is Com Task Force Four, Com Task Force Four. At seventeen-hundred hours this date, Japan sued for peace. This is official. The war is over. Japan has surrendered."*

A steam whistle blew, a deep, full throated roar that reverberated across the water. The sound seemed to come from one of the carriers.

And then a heavy anti-aircraft machine gun opened up with it's the tracer rounds arcing across the sky. It was followed by another, and the whistle-blast from a heavy cruiser.

"What the Hell's going on?" the Exec exclaimed.[32]

"Darned if I know," I said. "Tell the radio shack to come up on the harbor frequency. I don't care if there is radio silence. We better find out in a hurry!"

We didn't have long to wait.

A twenty-four inch search light began stabbing the air. More whistles blew, tracer bullets began to criss cross through the sky.

---

[32] In the U.S. Navy a ship's Executive Officer, XO, is responsible for personnel and functions as second-in-command.

And then the TB speaker on the bridge boomed into life.

"This is Com Task Force Four, Com Task Force Four. At seventeen hundred hours this date, Japan sued for peace. This is official. The war is over. Japan has surrendered."

A mighty roar went up from every ship.

The stomping feet, the hoarse cries of the men reverberated across the Eniwetok Atoll. We illuminated our ship as though we were an oceanliner. No more blackout curtains, no more working under darkened ship conditions.

We turned on our navigational lights for the sheer joy of seeing a red and green running light again. Across the lagoon it became a sustained pyrotechnic display of light and sound and at its' peak men kissed the steel decks with tears running down their cheeks.

The Exec stood in the lighted wheelhouse leaning on the whistle cord.

"Have you *ever* see anything like it!" he cried. "We'll remember it as long as we live!"

Long after the air banks were empty he kept pulling on the cord. A bead of moisture still clung to his chin.

"Hear that! We can go home now, Skipper," he kept saying, unable to contain his excitement—

"We can go home."

> *"Hear that! We can go home now, Skipper," he kept saying, unable to contain his excitement— "We can go home."*

REPORT OF SEPARATION FROM THE ARMED FORCES OF THE UNITED STATES

CHARACTER OF SEPARATION: HONORABLE
DEPARTMENT: U.S. NAVY

1. LAST NAME — FIRST NAME — MIDDLE NAME	BROWN Donald Wynn
2. SERVICE NUMBER	380 87 07
3. GRADE — RATE — RANK AND DATE OF APPOINTMENT	QMQC 8-1-43
4. COMPONENT AND BRANCH OR CLASS	USNR F-6

5. QUALIFICATIONS
- SPECIALTY NUMBER OR SYMBOL: QM-0201-49
- RELATED CIVILIAN OCCUPATION AND D.O.T. NUMBER: Mate Second 0-88.03

6. EFFECTIVE DATE OF SEPARATION: 30 MAR 54
7. TYPE OF SEPARATION: TRANSFERRED TO

8. REASON AND AUTHORITY FOR SEPARATION: NavPers 631 of 7-24-53 and Art. C-10323 BuPers Man.
9. PLACE OF SEPARATION: U.S. NAVPEC STA, L.BEACH, CALIF.

10. DATE OF BIRTH: 29 OCT 13
11. PLACE OF BIRTH: Little Rock, Arkansas
12. DESCRIPTION: SEX M, RACE CAU, COLOR HAIR Brown, COLOR EYES Blue, HEIGHT 72", WEIGHT 178

13. REGISTERED: NO (X)
15. INDUCTED: —

16. ENLISTED IN OR TRANSFERRED TO A RESERVE COMPONENT: YES (X)
- COMPONENT AND BRANCH OR CLASS: USNR F-6
- COGNIZANT DISTRICT OR AREA COMMAND: 11th ND, San Diego, California

17. MEANS OF ENTRY OTHER THAN BY INDUCTION: REENLISTED (X)
18. GRADE — RATE OR RANK AT TIME OF ENTRY INTO ACTIVE SERVICE: QMC

19. DATE AND PLACE OF ENTRY INTO ACTIVE SERVICE: 23 JAN 51, Yokosuka, Japan
20. HOME ADDRESS AT TIME OF ENTRY INTO ACTIVE SERVICE: Santa Ana, Orange, California

STATEMENT OF SERVICE FOR PAY PURPOSES

	A. YEARS	B. MONTHS	C. DAYS
21. NET NAVAL SERVICE COMPLETED FOR PAY PURPOSES EXCLUDING THIS PERIOD	16	04	02
22. NET SERVICE COMPLETED FOR PAY PURPOSES THIS PERIOD	03	02	08
23. OTHER SERVICE (Act of 16 June 1942 as amended) COMPLETED FOR PAY PURPOSES	00	00	00
24. TOTAL NET SERVICE COMPLETED FOR PAY PURPOSES	19	06	10

26. FOREIGN AND/OR SEA SERVICE: YEARS 03

27. DECORATIONS, MEDALS, BADGES, COMMENDATIONS, CITATIONS AND CAMPAIGN RIBBONS AWARDED OR AUTHORIZED:
Korean Service Medal (6 stars); National Defense Service Medal, United Nations Service Medal

28. MOST SIGNIFICANT DUTY ASSIGNMENT: USS LST-1123
29. WOUNDS RECEIVED AS A RESULT OF ACTION WITH ENEMY FORCES: none

32A. KIND & AMT. OF INSURANCE & MTHLY. PREMIUM: "G" $3.58
32B. ACTIVE SERVICE PRIOR TO 26 APRIL 1951: YES (X)
33. MONTH ALLOTMENT DISCONTINUED: TO REMAIN IN EFFECT

35. TOTAL PAYMENT UPON SEPARATION: $174.18
36. TRAVEL OR MILEAGE ALLOWANCE INCLUDED IN TOTAL PAYMENT: None
37. DISBURSING OFFICER'S NAME AND SYMBOL NUMBER: J.A. SCHERNS LT SC USN 535379

38. REMARKS: RECOMMENDED FOR REENLISTMENT

39. SIGNATURE OF OFFICER AUTHORIZED TO SIGN:
NAME, GRADE AND TITLE (Typed): H. I. WELLS, CH CLK, USN ASST PERS OFF

42. MAIN CIVILIAN OCCUPATION: Student

44. UNITED STATES CITIZEN: YES (X)
45. MARITAL STATUS: Married
46. NON-SERVICE EDUCATION: GRAM-MAR 8, HIGH SCHOOL 4

47. PERMANENT ADDRESS FOR MAILING PURPOSES AFTER SEPARATION: 449 Thalia Street, Laguna Beach, California
48. SIGNATURE OF PERSON BEING SEPARATED: Donald Wynn Brown

DD FORM 214 (1 JUL 52) — EDITION OF 1 JAN 50 IS OBSOLETE — INDIVIDUAL'S COPY (TO BE DELIVERED TO THE INDIVIDUAL BEING SEPARATED)

OCTOBER– NOVEMBER 1945

It wasn't that easy.

It was "curtain time" now. Time to remove the props of war and clean up the stage.

We blew up everything we couldn't salvage, and the remainder we dumped in the sea.[33]

And eventually, we returned to Manila.

It had become a staging area for used jeeps, bulldozers, trucks, canned sea rations, diesel engines, you name it, Uncle Sam had it at bargain-basement prices and Chiang Kai-Chek was our best customer.

While the barges were loading I lost half my crew under the wartime military point-system. Most of them were reserves and all good men, but when they hit the magic 24 points they took their walking papers and headed for home. Enough of them, and *just* enough, signed over into the regulars and some volunteered to stay on. They wanted to see China.

It wasn't so bad, really. We weren't under combat conditions any longer and we went at our work free and easy.

When we received our sailing orders a numb feeling crept over me, still I could feel a strange exhilaration. I had expected orders to the nearest port on the China coast but they read Tsingtao (Qingdao). The word seemed to leap from the pages in large black print.

---

[33] Captured Japanese Imperial Navy vessels were rounded up by the Navy, anchored in Bikini Atoll's lagoon, and used as targets. U.S. nuclear testing starting in 1946. Vast amounts of America's staged wartime supplies and material were simply discarded by being dumped overboard from ships, rather than being shipped back to the United States.

> *I looked in the mirror. Jesus! What had happened to me? My hair was almost grey. My face was lined.*

I went to my cabin took a shower and began to shave. My hand was shaking uncontrollably. I looked in the mirror. Jesus! What had happened to me?

My hair was almost grey.

My face was lined.

It had a tense look and then the past six years flooded in on me.

The North Atlantic, the torpedoings, the corpses floating in their life jackets in pools of oil. Frozen bodies on life rafts, the sea birds picking out their eyeballs. The flaming beach heads the scream of a ricoshaying shell. Cadavers lying on the beaches, eyes staring into the blinding sun.

Heroes every one of them.

The months and years of tension-filled sea watches had almost pulled me down. And then I considered the skippers on the cruisers and carriers. They really had something to worry about and I was thankful I only had my tug.

I finished shaving with both hands.

This was nonsense! I told myself. It's Tsingtao (Qingdao). You're apprehensive about what you'll find.

It was a colorful trip, up the west coast of Luzon, across the straits, then Northward along the west coast of Formosa. We had two barges on the end of our towing wire when we left Manila.

They were loaded to the Plimsol Lines with gear for the Chinese North Army and the naval base.[34]

The Navy, with a detachment of U.S. Marines, was moving back in.

The craggy peaks of the island thrust up through misty clouds. Off her rocky shores, fleets of sailing junks lay fishing.

We passed Shanghai and the Yangtze Delta, fifty miles at sea, it's a shallow desolate stretch of water, thick with yellow mud from the heart of China, with swirling eddies cluttered with debris.

We were on the last leg of the trip now with six years and a war between. Medical authorities say the body replaces all of its cells in seven years. At thirty-two, I was almost a different man, returning now as captain of my own vessel.

A mile off the old German Fort we brought the barges alongside then entered the harbor. At the far end of the bay a few textile mills were blown down but the warehouse was still there. I could see the spot where I had kissed Li Teh goodbye. I swung my binoculars up the hill. The apartment was still standing just as I had remembered it. We cleared with the Chinese Port Director, customs and health officers, made arrangements for offloading the barges and then I went below.

My hands were steady as I shaved. I put on my blues with a Lieutenant's two gold stripes on the sleeve, took a deep breath and stepped out of my cabin and off the ship.

I strolled ashore, waving aside the rickshaw coolies.

With the Japanese gone, the city was bustling as though they had never existed.

Walking up the hill I could see a white curtain in the window above Commadore's store. The landlord was still there, standing behind his counter as I had remembered him.

He looked the same, a little thinner perhaps, but his dark skin was ageless and he could still smile. His eldest son was with him, grown now, with a man's voice and there were a couple of new young ones running around.

---

34 Plimsoll lines indicate reference marks typically located on or near the ship's bow on the hull. These marks show the maximum depth to which the vessel may be safely immersed when loaded with cargo.

He didn't recognize me, bowing me into the store and rubbing his hands as though I were a foreign devil customer.

"Have I changed that much, Wong?" I asked.

"Ai—EE" he exclaimed. "My eyes must be getting poor." Calling to his wife he threw his arms around me and we stood there hugging and laughing with memories. He pulled me into the living room while Number One Tai Tai put on a pot of tea. After we had simmered down a little I asked, "Where can I find Li Teh?"

Wong's face wrinkled, his lips twisted and then he broke into tears. His wife followed suite, wringing her hands and wailing. Sorrow in a Chinese wells up from the very core.

He wiped his eyes and murmured, "Deng-Bu-Hau! Dead! All Dead!" He counted on his fingers and moaned, "Wan, Wan, Yi-bai-wan!-Thousands, millions of them."

"It's been the same all over the world," I said. "Its okay, Chu Cheng. It's all over with."

He looked at me uncomprehendingly and then I added, "How about Doctor Graham? Is he dead?"

Wong shook his head. "Only half-dead."

And then he broke down again. In the silence we could hear the cry of a vendor, the squeaking wheel of a heavy cart going down the hill.

I gave his eldest son a handful of *mex*. "A little food perhaps," I suggested. "Some of those ripe eggs he likes so well, I must know what happened."

Gradually the story unfolded. We sat over a pot of h ot rice wine for hours and as he got into it, his eye lids were half closed, seeing nothing, feeling for the words that were in his mind and his heart.

His face smoothed out as he talked, as though his words were releasing him from the clutch of a mystic dragon.

His words told a story of evil and pain and sacrifice and beauty, but a beauty formed of the unconquerable spirit existing in the human heart, this energized flux of compassion that has existed in mankind through the ages.

--30--

U. S. NAVAL RECEIVING STATION
U. S. NAVAL STATION
LONG BEACH 2, CALIFORNIA

NM17/P19-5
RS4/HW:se

30 MAR 1954

From: Commanding Officer, U. S. Naval Receiving Station,
U. S. Naval Station, Long Beach 2, California

To: BROWN, Donald Wynn, 380 87 07, QMQQ, USNR F6

Subj: Release from active duty and transfer to inactive duty in Class ...F-6... Fleet Reserve

Ref: (a) NavPers 631 of 7-24-53 & Art. C-10223 BuPers Manual

1. Examined and found physically qualified for release from active duty.

   W. W. WATERS CDR MC USN   Medical Officer

2. Detached 30 MARCH 1954. Proceed to your home. You will regard yourself released from all active duty effective at 2400 this date at which time you are transferred to inactive duty in Class F-6, Fleet Reserve.

3. You have stated that upon release from active duty your "present address", the place at which you may be reached at any time by orders or other official communications, will be 449 Tealia Street, Long Beach, California. You may change your "present address" at any time but such change shall be reported promptly by letter to your commanding officer and to the Bureau of Supplies and Accounts, Field Branch (Special Payments Division), Cleveland 15, Ohio, in order that your retainer pay checks and official communications to you may not be delayed. Such notification shall include the same information prescribed in the following paragraph and, in addition, your old address as well as your new address.

4. On the basis of the "present address" set forth above, your commanding officer will be the Commandant, 11th Naval District, whose address is San Diego, California. Your service record will be maintained by that command and any questions regarding your status in the Fleet Reserve should be addressed to him and should include your full name, rate, service number, Fleet Reserve Class, and "present address."

5. You will keep yourself in readiness to respond to orders to active duty in time of war or national emergency. You are advised that in time of peace you may be required to perform not more than 2 months active duty in each 4-year period. While on inactive duty you will be examined physically at least once during each 4-year period. Inform your commanding officer of any change in health which might prevent active service in time of war. You shall promptly answer all official correspondence addressed to you as such and shall comply with instructions contained therein. While on inactive duty you are subject at all times to the laws, regulations, and orders for the government of the Navy.

6. Permission to leave the United States for periods in excess of 30 days in time of war must be requested from the Chief of Naval Personnel, Department of the Navy, Washington 25, D. C., via your commanding officer. In time of peace, such permission must be requested from your commanding officer. You may not, however, visit belligerent countries at any time without prior approval of the Chief of Naval Personnel.

NM17/P19-5

Disbursing Officer (2)
Medical Officer (1)

NADO, NS, LONG BEACH, CALIF. 3/30/54. No mileage allowance paid on initial transfer to inactive fleet reserve F-6. Travel allowance is payable in accordance with paragraph 4155 (case 12) of Joint Travel Regulations.

NADO, NS, LONG BEACH, CALIF. 3/30/54. No mileage allowance paid on initial transfer to inactive fleet reserve F-6. Travel allowance is payable in accordance with paragraph 4155 (case 12) of Joint Travel Regulations.

H. K. LAWRENCE, PAYCT USN
for and in the authorized absence of
J. A. SCHIERES LT SC USN

11ND Form 2619-C (New 11-52)

AFTERWORD

Against Time's sweeping relentless passage, Life's only defense is renewal.

World War II, whose devastating proportions re-aligned America and the world's destiny, came if not to an end in 1945, at least to a reductive pause.

Conflicts of various political alignments and economic nature marked the following years, and their often cruel adjustments would continue in the decades to come, coursing like the rip currents Don Brown learned to both use and avoid as a young Laguna Beach lifeguard.

Young Don's making good on his promise to return to China, was tempered upon his arrival by his learning of the fate of the two pivotal people who had prompted his dutifully kept promise.

It is important to understand that this wasn't just a young man's idealized romantic impulse. His firm commitment had been resolutely demonstrated when he may have taken another man's life at close and violently brutal quarters in personal revenge at a time——since America was not then at war with Japan——might well have triggered a precipitating international incident.

Had Don been caught under existing U.S. military law and indicted before a U.S. Navy court-martial, given the political tenure of that time, the result could have been the end of his naval career and long imprisonment, not to mention his certain execution, had he been apprehended by the Japanese.

War, as it had for so many others, swept Li Teh away. Suffering and deprivation terminally reduced Dr. Grahm's remaining life. And with them was swept Don's anticipation of joy at their survival.

Also lost was some hoped-for future they both represented for him. As Don had put it to me after reading that passage:

"I felt like the war I'd fought had personal side——so that Li Teh could live and Doctor Graham could continue his work and his Mission. That hadn't happened. I'd left too soon and returned too late."

Nevertheless, Don Brown was not a man to ignore or forget a promise. He kept them alive in memory.

Upon his discharge, Don would return to Laguna Beach, marry and start a family. Still, as with so many other veterans, along with his medals, ribbons and commendations, he carried the war's impact with him, a life-long emotional shadow of the period's accumulated losses——his own, as well as those of his cohorts with whom he'd served.

This would form the impetus guiding his writing, his slice of near-history in tribute to their memory, to his love, respect, and dutiful sense of honor for those for whom he deeply cared.

Throughout Don's life he effortlessly expressed a deep level of caring and compassion for others. It was a remarkable quality that marked him as, if not a saint——as he'd readily admit——but a wonderful human being whose energy, integrity and creative vitality are the soul of this narrative, and dignify his life.

Craig Lockwood

--30--

A former Laguna Beach Lifeguard, Lockwood worked on-staff for his hometown newspapers, the Laguna Beach Post, the South Coast News, and the seven-edition county-wide daily, The Daily Pilot and Times/Mirror as a reporter/columnist, and multi-magazine conglomerate, Peterson Publications as an editor.

Lockwood is a lifetime member of the US Marine Corps Combat Correspondent's Association, Society of Professional Journalist, American Society of Journalists and Authors, and the Orange County Press Club.

www.ingramcontent.com/pod-product-compliance
Lightning Source LLC
Chambersburg PA
CBHW042138290426
44110CB00002B/48